Fundamentals of LIFE SCIENCE

Second Edition

LAB BOOK FOR BIOLOGY 189 AT NEVADA STATE COLLEGE

Nevada State College

Edited by
Nathan K. Silva, Ph.D. ▪ Cheryl Vanier, Ph.D.
Vikash Patel, MPH, MSC

Cover image © Shutterstock, Inc.

Kendall Hunt
publishing company

www.kendallhunt.com
Send all inquiries to:
4050 Westmark Drive
Dubuque, IA 52004-1840

Copyright © 2013, 2016 by Nevada State College

ISBN 978-1-4652-9734-1

Kendall Hunt Publishing Company has the exclusive rights to reproduce this work,
to prepare derivative works from this work, to publicly distribute this work,
to publicly perform this work and to publicly display this work.

All rights reserved. No part of this publication may be reproduced,
stored in a retrieval system, or transmitted, in any form or by any means,
electronic, mechanical, photocopying, recording, or otherwise,
without the prior written permission of the copyright owner.

Printed in the United States of America

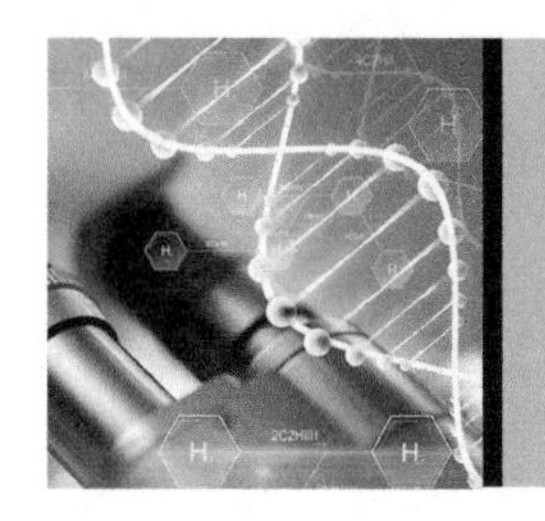

Contents

LAB 1	**Experiments and Graphing**	**1**
LAB 2	**Measurements and Metrics**	**9**
LAB 3	**Microscope Usage**	**17**
LAB 4	**Osmosis and Diffusion**	**25**
LAB 5	**Enzymes in Digestion**	**31**
LAB 6	**Plant Stomata**	**39**
LAB 7	**Photosynthetic Pigments**	**45**
LAB 8	**Mendelian Genetics and Probability**	**55**
LAB 9	**DNA Extraction from Strawberries**	**65**
LAB 10	**The Central Dogma of Genetics**	**73**
LAB 11	**Bird Beak Adaptation**	**85**
LAB 12	**Biochemical Evidence for Evolution**	**95**
LAB 13	**Human Blood Typing**	**101**
LAB 14	**Physiology of the Circulatory System**	**109**
LAB 15	**Simulated Disease Transmission**	**125**
LAB 16	**Fetal Pig Dissection Instructions**	**131**

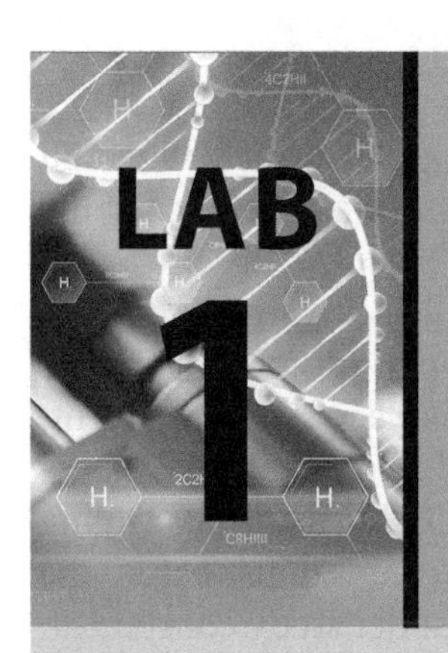

Experiments and Graphing

INTRODUCTION

A well-designed experiment is the cornerstone of the scientific method. In this lab, we will be exploring the parts of a good experiment and how results can be displayed in graphical format. As you will see, it is very helpful to understand the experiment if you want to graph it properly.

Experiments

In a simple experiment, we usually want to know what will happen if we expose our test subjects to some treatment. The test subjects can be organisms, such as lab mice, bacteria, and humans, or the test subjects can be non-living things, such as rocks or pieces of DNA. As the person designing the experiment, you get to decide what the test subjects will be, and what treatment you will expose the test subjects to. The more test subjects you involve in your study, the more confidence you will be able to have in the results.

Because we need information about what changed when we added a particular treatment, some test subjects are assigned to one or more treatments chosen by the experimenter, and the remaining test subjects are assigned to a control treatment. You make a control treatment by doing exactly the same procedure as for the treatment you have devised, but you stop short of delivering the actual treatment. For example, if someone was studying how well removing a certain type of tumor eased cancer symptoms, some test subjects would be assigned to have their tumor surgically removed, while others would be assigned to a control group. The people assigned to the control group would receive the same anesthesia and surgery, but the surgeon would not remove the tumor once he or she was inside the body. In a different example, if you wanted to know the effect of hotter temperatures on bacterial growth rates, you would assign some bacteria to be placed on a warming plate that was turned on to heat them.

© Syda Productions/Shutterstock.com

The control bacteria would be placed on the same kind of warming plate, but you would not turn it on, so the bacteria would be handled in exactly the same way, except for the temperature.

Any quantity that changes or varies in an experiment is called a variable. In the previously mentioned bacteria experiment, we had two temperatures: room temperature and warmer than room temperature. Temperature is therefore a variable in this experiment. Any quantity that is not allowed to change in an experiment is called a constant. In the bacterial growth rate example, there are many constants that we can identify, such as the warming plate, the room, the type of bacteria, and the food that the bacteria received just before the study. A good experiment will make everything except the variable under study constant so the results clearly show the effect of the variable alone, without complicating factors.

So far, we have been focusing on a variable chosen by the experimenter. We call the variable or treatment chosen by the person designing the study the independent variable because it does not depend on anything else in the study. It is part of the study design. After the treatment is applied, observations are made on the test subjects. Because these observations are not all going to be the same across test subjects, they will also form a variable in the experiment. In the example where the bacteria were grown in different temperatures, you might measure the number of times the bacteria divided in a particular period of time to assess growth. Growth, however, is a different kind of variable than temperature in this study. They are both variables, but temperature was chosen by the experimenter, whereas we want to know if growth changed in response to temperature. Because we want to know if it depends on another variable in the experiment, growth is a dependent variable. The dependent variable is what you measure during the course of the study after the treatment is applied.

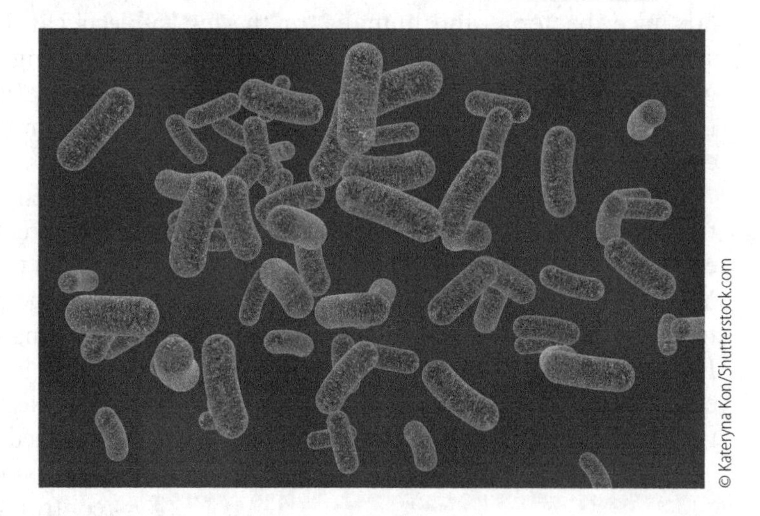

© Kateryna Kon/Shutterstock.com

Types of data that make up variables

Whether dependent or independent, we can choose to express variables in different ways. When we think of data, we often think of numbers. Numbers express quantities, so numeric data is called quantitative data. Some variables, such as color, are not easily measured as numbers. In that case, test subjects may be assigned to categories. This type of data is qualitative data. Qualitative means that there is no reasonable way of ordering the values from greatest to least. There is an intermediate type of data that lies somewhere between quantitative and qualitative. This is when we have categories (which usually means qualitative data) that can be ordered in some sensible way (which is generally a feature of quantitative data). We can call this type of data quantitative categorical to distinguish it from purely numeric data (which we could call quantitative continuous data) and purely qualitative data. In the example with the bacteria, if we measured the number of times the bacteria divided, this would be numeric, so it would be quantitative continuous data. If we grouped our bacteria into small, medium, and large classes, then we have groups that can be ordered, so this would be quantitative categorical data. If we decided to observe the bacteria and group them as rough, smooth, bumpy, or ciliated (hairy), we would be working with qualitative data.

Plotting variables in graphs

If you have very few values, or if the dependent variable is categorical, you may need to show the results in a table. Otherwise, you will want to use a graph. The type of data in the variables will determine the graph that is most appropriate for showing the outcome of a study (Table 1).

Table 1—Types of data required for each type of graph, including special requirements for each. Note that the 'bar' graph is called a 'column' graph in Microsoft Excel

Graph Type	Independent Variable	Dependent Variable	Special Requirements
Scatter	Quantitative continuous	Quantitative continuous	Connecting dots is not sensible
Line	Quantitative continuous	Quantitative continuous	Connecting dots is sensible
Bar	Qualitative or Quantitative categorical	Quantitative continuous	Data must be summarized as one value per catogory
Pie	Qualitative or Quantitative categorical	Quantitative continuous	Data must be summarized as one value per catogory and sensibly expressed as a proportion relative to the other categories

In the bacteria example, the independent variable, temperature, was expressed as qualitative or quantitative categorical data, so a scatter or line plot would not work for this study design. To decide between a bar and pie graph, we have to look at the special requirements. If we chose to measure the dependent variable, growth, as number of times the bacteria divided, we would have a list of numbers. We must have one value for each of our two temperature groups, so we could average all of the numbers for each group. Let's imagine that we calculated that the bacteria divided 2.5 times on average in the room temperature treatment and 2.9 times on average in the warm treatment. The pie graph does not really make sense in this context because it requires that these two average values be summed and expressed as percentages. It would not be very clear to a reader to show this information as a pie graph! So we are left with a bar graph as the best option.

Important features of a graph

Whenever you prepare a graph, there are certain conventions that can help you produce an informative and clear product. Once you have chosen the type of graph, you need to consider the following features:

- Which variable goes on the *x* (or horizontal) axis? By convention, the independent variable will always be placed on the *x* axis and the dependent variable will be on the *y* (or vertical) axis.

- Values on the axes: If you have categories, label each group with the appropriate category name. If you have numeric data, first find the highest and lowest values in your data set. Use those numbers to determine a reasonable interval between tick marks for that axis. For example, if your values run between one and nine, it would be reasonable to have values between zero and nine, counting by ones (ten tick marks). If your numbers spanned from 3 to 99, then having tick marks for values from 0 to 100, counting by tens, would work nicely. The numeric value between tick marks should always be consistent from one interval to the next; you should never have, for example, one tick mark labeled '5,' then next '10,' and then '25.' The interval between the first two is 5, whereas the interval between the latter two is 15. Inconsistent intervals produce misleading graphs.
- Label and units for each axis: Both axes must be labeled so that a person who doesn't know about your data can understand the graph. If the value was measured in units, this should also appear on the axis label.
- Caption: By convention, figure captions must be placed below the graph. To make it easy to refer to the figure in your text, give the graph a number, with the first graph in the paper labeled as 'Figure 1,' the second as 'Figure 2,' etc. Following the label should be a brief description of what information is in the graph. Note that this is not the place to interpret the pattern in the data; you are simply trying to let a reader know what data the graph is depicting.

Examples of graphs and table

In our bacteria example, we had determined that the appropriate graph type was a bar graph. We will use the average values to represent each group.

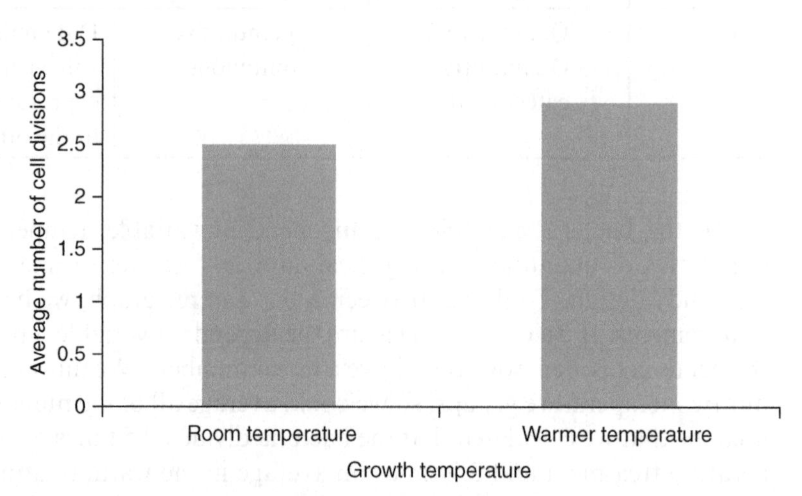

Figure 1 Average number of cell divisions in bacteria after 48 hours of growth at two temperatures

There are examples of the other graph types using a similar experimental setup. Notice the different types of data that were used in each case.

Scatter plot: If we measured bacterial growth for two samples across a range of temperatures, a scatter plot may be most appropriate.

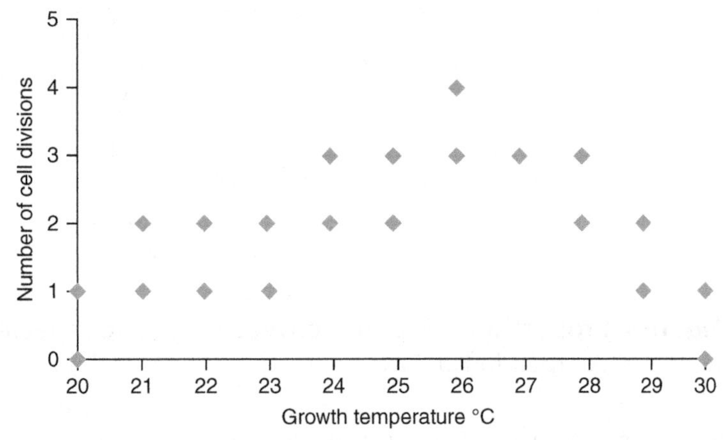

Figure 2 Number of cell divisions in 48 hours in bacteria grown across a range of temperatures

Line plot: If we wanted to look at two different species of bacteria across a range of temperatures, a line plot may be the best option.

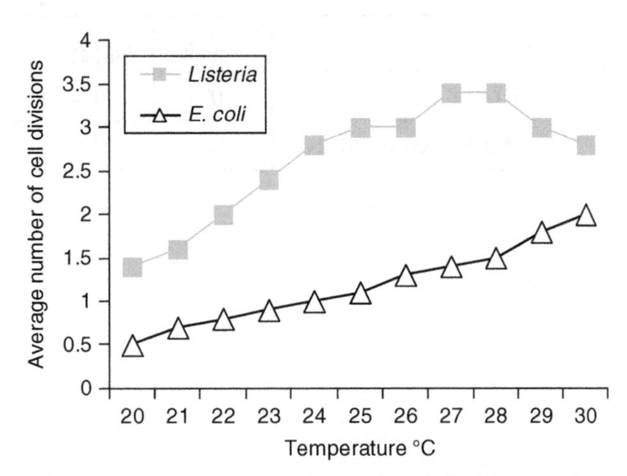

Figure 3 Average number of cell divisions after 48 hours for two types of bacteria across a temperature gradient

Pie plot: If we wanted to know how these growth rates translated to a situation where the two bacteria were grown together at a certain temperature, we might grow them on the same plate until the plate was completely covered, then show the proportion of space that each took over in a pie plot.

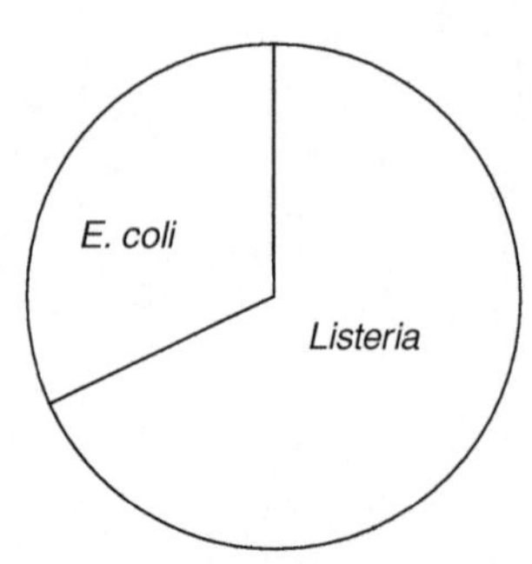

Figure 4 Proportion of plate covered by two species of bacteria grown in competition at 25°C.

Table: If we had categorized the bacteria after temperature treatment as rough, smooth, bumpy, or ciliated (hairy), we may prefer to show the data in a table. A table should also have labels for every column and, if appropriate, row. The caption for a table sits above the table, and it is numbered (for example, 'Table 1' and 'Table 2').

Table 2—Percent of bacteria with particular textures after being grown for 48 hours at different temperatures

	Rough	Smooth	Bumpy	Ciliated
Room Temperature	10%	70%	10%	10%
Warm Temperature	30%	20%	40%	10%

Name ______________________________ Date ______________

Measurements and Metrics Lab—Student Worksheet

QUESTIONS

Josie wanted to study the effects of a high-fat diet on rats. She had 20 rats that had all been born on the same day, were raised in individual cages in her lab, and who were all of the same breed. She assigned 10 of them to eat a high-fat diet and 10 to eat the usual rat chow fed to rats in her laboratory. After 30 days on each diet, she measured running speed in a maze. Answer the following questions regarding her study:

© Vasiliy Koval/Shutterstock.com

1. What are the test subjects?

2. What is the independent variable?

3. What is the dependent variable?

4. What is the control?

5. Name two constants in Josie's study.

6. In Josie's study above, what type of data is the diet?

7. What type of data will the running speed be?

8. What type of graph would work best for this data set?

9. If the average running speed of high-fat rats was 2 km/hr and the average running speed of low-fat rats was 3 km/hr, draw a graph to show the experimental results. Be sure to include all of the important elements of graphs.

10. If Josie had decided to assign sets of rats to diets ranging from 5% to 60% fat, moving up in increments of 5%, what graph type would have been most appropriate for her data?

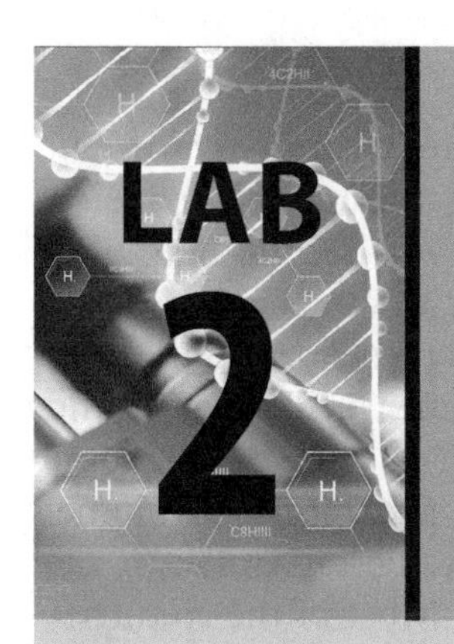

Measurements and Metrics

BACKGROUND

A patient comes into a hospital for an inpatient procedure. She is under anesthesia, and her blood pH has to be maintained between 7.35 to 7.45 for her body to function properly. If her pH goes below 7.35, then the doctor may prescribe sodium bicarbonate, which is a buffer to help stabilize the blood pH within the range mentioned. If you were the nurse, the doctor may instruct you to intravenously infuse 50 milliequivalents (mEq) of sodium bicarbonate in a specific glucose solution at a rate of 1 L/hour. Although it may be daunting to understand at this point what the unit milliequivalent is and how it relates to liters, this lab activity will help us understand some of the basics of how measurements are applied. Understanding measurements can be vital for the survival of patients.

A big part of science is communicating measurements and data. This communication of information requires there to be consistent units of measurement. In the past, one unit of measurement that was used was the cubit. The cubit is defined as the distance between the top of your middle finger and your elbow. This was a convenient measurement because as long as you had an arm, you had something that you could use to make the measurement, but it was an inconvenient measure because everyone does not have arms of exactly the same length. Let's say that you are building a house and you need a ceiling beam that is exactly 15 cubits long. If you send someone else out to pick up that piece of lumber, you had better hope that their arms are exactly the same length as your arms, or the piece of lumber will not be the correct size. Hopefully this shows why we need to use a unit of measure that will be the same no matter who makes that measurement. An inch is an inch for everyone; the same is true for meters, gallons, or seconds.

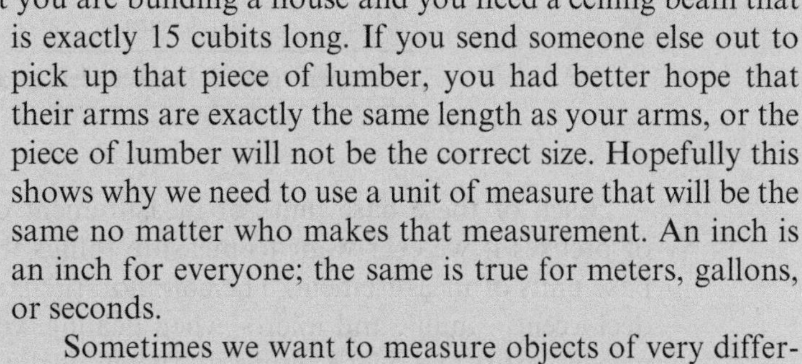

© Everett Collection, 2013. Used under license from Shutterstock, Inc.

Sometimes we want to measure objects of very different sizes. Assuming you are most familiar with the United States customary system or British Imperial units, if you wanted to measure the height of a door, you would likely use feet. If you wanted to measure the length of a pencil, you would likely use inches. If you wanted to measure the distance between Los Angeles and Las Vegas, you would likely use miles. Some of these conversions we are very familiar with, we know that there are 12 inches in a foot and three feet in a yard, but how many feet are there in a mile? The answer is 5280. That is a tougher conversion to remember. Then if I ask how many yards there are in a mile, you need to take 5280 and divide it by three (1760 yards per mile). For most people, these conversions are not automatic, these conversions don't just "make sense"; it takes memorization. When you take into account that these are just the conversions for measuring length and there are many other conversions to memorize for volume and mass, we would assume that there has to be an easier way to measure objects of varying size, mass, or volume. And there is. This simpler system is called the International System of Units, or the SI system, and it is the most modern form of the metric system. Most of the countries in the developed world use this SI system. There are only a few exceptions, the United States and England being some of those. Science and scientific discovery is an international endeavor and it is especially important to convey measurement and data in a way that is consistent. So science has adopted the SI system as the sole system of measurement used to record and communicate scientific data. In this class and in future science classes, you will need to become familiar with the metric system and the way that this system works. Fortunately, the metric system is organized to limit the amount of memorization you will need to do. You first need to learn base units for each type of measurement, then the base units will be combined with a consistent set of prefixes to change the size of the unit as needed.

"What are the base units of measurement?" you may ask. The base unit of measurement for length in the metric system is the meter. The base unit for measuring volume is the liter. The base unit for measuring time is the second. The base unit for measuring mass is the gram and the base unit for measuring temperature is degrees centigrade (Celsius).

Measurement	Unit	Symbol
Length	Meter	m
Volume	Liter	L
Time	Second	s
Mass	Gram	g
Temperature	Degree Centigrade (Celsius)	°C

Each of these base units of measurement can be modified using a consistent set of prefixes if we ever want to measure things that are vastly different in size than the base units of measurement. The common prefixes that we will be using in this class are deci-, centi-, milli-, and micro- when dealing with things smaller than the base unit of measurement and Mega- and kilo- when dealing with things larger than the base unit of measure. We will first deal with the prefixes for making units of measure smaller than the base unit. The prefix deci- means 1/10th or 10 times smaller than the base unit of measure. One base unit is equal to 10 deciunits. The next prefix is centi-, centi- means 1/100th or 100 times smaller than the base unit of measure. One base unit is equal to 100 centiunits. Milli-means 1/1,000th and micro- means 1/1,000,000th. The common prefixes for things that are larger than the base unit of measure are kilo- and Mega-, kilo- means 1,000 times larger than the base unit of measure and Mega- means 1,000,000 times larger.

© Kimberly Hall, 2013. Used under license from Shutterstock, Inc.

Prefix	Comparison to the Base Unit of Measure
Mega- = M	1,000,000 times larger (10^6)
kilo- = k	1,000 times larger (10^3)
deci- = d	1/10 or 10 times smaller (10^{-1})
centi- = c	1/100 or 100 times smaller (10^{-2})
milli- = m	1/1,000 or 1,000 times smaller (10^{-3})
micro- = μ	1/1,000,000 or 1,000,000 times smaller (10^{-6})

By learning these six prefixes, you can measure things of vastly different sizes, masses, lengths, and times, and convert between those prefixes easily. There are several methods of converting from one unit to another. We will introduce two in this introduction. Use whichever method makes the most sense to you. If you learned a different method in the past and feel comfortable with it, use that method to convert units. One way to easily convert from one prefix to another is through the use of a number line and the sliding decimal method of unit conversion.

Mega-			kilo-			Base	deci-	centi-	milli-			micro-
M			k			__	d	c	m			μ
0	0	0	0	0	0	0.	0	0	0	0	0	0

When you are given a particular measurement; for example, 25 meters, and are asked to convert it into decimeters, there are a few questions that we need to ask ourselves before we do this conversion. The first question to ask is whether we are converting to a larger unit of measure or a smaller one. In this case decimeters are smaller than meters, so we are converting into a smaller unit of measure. Whenever converting to smaller units of measure, we need to move the decimal point a certain number of spaces to the right. When we are converting to larger units of measure, we need to move the decimal point a certain number of spaces to the left. "Where is the decimal point?" you may ask. If the decimal point is not explicitly shown in the number, then is it always to the right of the last digit that you are given. In our example of 25 meters, the decimal point is to the right of the 5, so it is 25.0 meters. We want to move the decimal point on our number line from where it is to the right of the prefix we are converting to. We are starting at the base unit of measure (meters) and converting to decimeters, so we need to move that decimal point one spot to the right. 25 meters is equal to 250 decimeters. Let's say that we have 3200 microliters (μL) and we want to know how many centiliters (cL) that is. The decimal starts at the right of the last 0 in 3200, cL are larger than μL, so our decimal point is going to move a certain number of spaces to the left. Consulting our number line, we see there are four spaces that separate the right of micro- from the right of centi-, therefore 3200.0 μL is equal to 0.32 cL. We do not always need to start or end with a base unit of measure when we use this method (sliding decimal) of unit conversion.

Another common method for converting units is the ratio method. The key to using the ratio method is to use the fact that a unit that is both on the bottom and the top of a ratio is cancelled out; just as 5/5 = 1, meters/meters = 1. So if we wanted to convert 10,000 μm (micrometers, which we could also express as 10^4 μm) into meters, we could use the fact that 1 μm = 10^{-6} m:

$$\frac{10^4\ \mu\text{m}\ |\ 10^{-6}\text{m}}{|\ 1\ \mu\text{m}} = 10^{-2}\ \text{m, or 0.01 m (the } \mu\text{m cancelled out)}$$

(recall that when you multiply numbers with exponents, you simply add the exponent values together)

You can also do these in several steps, if you need to first get to the base unit in order to finish the conversion. For example, how many micrograms (μg) are in 5 kilograms?

$$\frac{5\ \text{kg}\ |\ 10^3\text{g}\ |\ 1\ \mu\text{g}}{|\ 1\ \text{kg}\ |\ 10^{-6}\text{g}} = 5 \times 10^9\ \mu\text{g, or 5,000,000,000 } \mu\text{g (the kg and g cancelled out)}$$

(recall that when you divide numbers with exponents, you change the sign of the exponents in the denominator when you add them together with the numbers in the numerator)

This method is extremely flexible, allowing you to use any unit equalities available to you to convert. An example of its flexibility is if we wanted to convert a length measurement from 10 miles into kilometers. If you knew that 2.54 inches = 1 cm, and that 12 inches = 1 foot and 5280 feet = 1mile, this can be accomplished in several steps:

$$\frac{10\ \text{miles}\ |\ 5280\ \text{feet}\ |\ 12\ \text{inches}\ |\ 2.54\ \text{cm}\ |\ 1\ \text{m}\ |\ 1\ \text{km}}{|\ 1\ \text{mile}\ |\ 1\ \text{foot}\ |\ 1\ \text{inch}\ |\ 100\ \text{cm}\ |\ 1{,}000\ \text{m}} = 16.1\ \text{km}$$

It doesn't matter how many steps it requires; the main thing is that you cancel out each unit and replace it with another which will get you closer to your goal. It is extremely important that the units work through to the correct quantity. When you a complete a conversion, you should also do a 'reasonableness test' based on your general sense of the quantities involved. For example, we generally know that a km is a bit smaller than a mile. That 10 miles would be 16.1 km is reasonable; if we have calculated that 10 miles is 10,000 km, we have clearly made a mistake.

Some measurements are actually a ratio between two or more measurements. Examples of this are density, velocity, and concentration. Density is a ratio of mass/volume, velocity is a ratio of distance/time and concentration is a ratio of the amount of solute/the amount of solvent. These ratios allow us to compare the properties of different substances, even when we have very different amounts of those substances. Why do ice cubes float in a glass of water? Because of the difference in density. We also know that 3 kg of sand would take up much less space than 3 kg of styrofoam. Why? These two substances have a different density, so even though in this example they have the same mass, they will occupy a different volume. We will be calculating the density of two different liquids in today's lab.

We will also explore sampling in this laboratory exercise. Sometimes there is a population that we are interested in, but to get an exact count would be time-consuming, inefficient, and/or impossible. Examples would be counting the number of trees in a forest, or the number of anchovies in a school or the number of phytoplankton in the ocean. Even when getting an exact count is not feasible, we are able to follow a few basic steps to get an accurate estimate of those values in a very time-efficient manner.

The key strategy is called sampling, which requires the following steps:

- take the entire study area and divide it into smaller subunits which can be more easily handled
- randomly select several of the subunits to count individuals in order to get an average of how many are found in each
- determine how many subunits (selected and unselected) make up the entire population
- multiply the number of subunits by the average number of organisms per subunit to estimate the total number of organisms in the population.

We will be seeing how this works as we estimate the number of bacterial colonies there are on a simulated bacterial plate.

PROTOCOLS

Activity 1: Calculating the density of oil and water

Materials needed for each lab group

2 Graduated medicine cups
1 Transfer Pipette
1 Large clear plastic cup
Oil
Water

1. With what you currently know about oil and water, which would you hypothesize to be more dense (have a higher density), oil or water, and why? Record your initial hypothesis in the "Questions" section of the Student Worksheet.
2. Obtain the two graduated medicine cups. Use those which are already oily for oil and those which are not oily for water.
3. Get a mass measurement of each cup while it is still empty using the scale and record the mass in grams in Data Table 1. (Note: You may need to lift the colored plastic dust cover in order to place the cup on the metal disk in order to get a mass measurement from the scale)
4. Using the lines on the cup, measure 30 mLs of oil into the oil cup, and 25 mLs of water into the water cup. (Note: Get as close as you can to your desired volume by pouring the liquid directly into the cup, then use the transfer pipets for the fine-tuning of your volume.)
5. Get a mass measurement of the filled cups and record the mass in grams in Data Table 1. (Note: Use the same scale as the first round for each of these measurements to remain consistent.)

6. Calculate the mass of your oil and water sample by subtracting the mass of the empty cup from the mass of the filled cup.
7. Calculate the density of each liquid by dividing the mass by the volume. (Note: The volume was what you measured using the lines on the cups. The units of density that you will be using for your measurements are (g/mL).)
8. After calculating the density for each liquid, compare the results to your original hypothesis. Now predict what you expect to happen when you mix both liquids together in the plastic cup. (Note: Be sure to record your prediction in the "Questions" section of this lab, before mixing the liquids.)
9. Test your prediction by combining the liquids in the large plastic cup and observing the results. Record your observations and write a statement comparing your prediction and the observed behavior of the combined liquids in the "Questions" section of the lab.
10. Clean up this activity by pouring the oil and water mixture down the sink and rinsing all of the cups with water and leaving them on the paper towels near the sink to dry.

Activity 2: Quick, efficient, and accurate estimations

Materials needed for each lab group

1 Simulated Bacterial Colony Plate
1 Gridded Transparent Overlay Sheet
1 Rule capable of measuring in centimeters

1. After obtaining the Simulated Bacterial Colony Plate, but before a detailed counting or calculation, each member of the lab group needs to make a guess as to the number of simulated colonies on the plate. Record your guess in the "Questions" section of this lab. (Hint: The number will be somewhere between 10 and 10,000 colonies.)
2. After each member has recorded their guess, place the gridded transparent overlays over the colony plate so that the larger circle is divided into several smaller squares. The entire plate represents the population you are trying to measure, and each square in the grid is a subunit of the entire plate.
3. Pick ten of these squares at random and count the number of simulated colonies in each square, recording these numbers in Data Table 2.
4. Calculate the average number of colonies per square by adding all ten of these values together, then dividing by ten.
5. Calculate the number of colonies per cm^2 by measuring the diameter of the circle, calculating the radius, then calculating the area of the circle. The equation for calculating the area of a circle is πr^2. For the purpose of this lab, feel free to use 3.14 as the value for π. The radius of a circle is equal to ½ the diameter. Be sure to square the radius before multiplying it by π.
6. Multiply the area of the circle by average number of bacteria per square to get the number of colonies you estimate in the entire population.
7. Clean up for this activity involves returning the Simulated Bacterial Colony Plate, Gridded Transparent Overlay, and Ruler to the location they were obtained.

Activity 3: Metric Conversions

1. In the "Questions" section of the lab, complete the metric conversions using either the sliding decimal method of unit conversion, or whichever method you prefer.

Name ____ Date ____________________

Measurements and Metrics Lab—Student Worksheet

QUESTIONS

1. Before performing the calculation, which do you think is more dense: oil or water? Why?

2. What is the density that you calculated for these two liquids?
   Density of Oil = __________ g/mL Density of Water = __________ g/mL
   How does this compare with your initial hypothesis?

3. What occurred when you combined the two liquids in a cup? Was that predictable from the density values you observed?

4. How would you expect the results to differ if you used twice as much oil as water? Would you still get the same general result? Why?

5. Pumice is a type of volcanic rock that floats when placed in water. What does this tell you about the density of pumice?

6. What is your initial guess as to the number of simulated bacterial colonies on the plate you have been provided? (no counting, just guess)

7. What was the average number of simulated bacterial colonies that you calculated for each square?

8. What is your calculated estimate for the number of simulated bacterial colonies on the entire plate and how does that number compare with your original guess? Which do you think is more accurate?

9. How many colonies would you expect in every cm^2 of the circle based on your calculations?

Name ______________________________ Date ______________

10. Why would a forester use a method similar to the one we used to calculate the number of trees in a 10,000 m^2 forest?

Complete the following metric conversions. Carefully note the distinction between M (Mega), m (milli), and μ (micro).

11. 25 cm = ________ dm
12. 100 kL = ________ ML
13. 1 s = ________ μs
14. 11.25 dg = ________ g
15. 36.7 g = ________ cg
16. 500 L = ________ kL
17. 3,000 μm = ________ cm
18. 10,000 s = ________ Ms
19. 567 dL = ________ kL
20. 8 kg = ________ g
21. 0.536 m = ________ cm
22. 0.45 ms = ________ cs

DATA TABLES

Data Table 1

	Oil	Water
Empty cup mass (g)		
Filled cup mass (g)		
Mass of liquid (filled-empty) (g)		
Volume (mL)		
Density (g/mL)		

Data Table 2

sample	1	2	3	4	5	6	7	8	9	10
# colonies										

Calculate the following. Be sure to include your units!

- Average number of colonies per square: _______
- Diameter of the circle: _________
- Radius of the circle: __________
- Area of the circle ($\pi * r^2$): __________
- Each square is one cm^2. How many squares are in the circle, based on your calculation of area? ____________________
- Calculated number of total colonies in the circle: __________

Microscope Usage

BACKGROUND

The limit of human vision means that much of the world of cellular biology is lost to us if we did not have some way to magnify these tiny objects. The discovery of bending rays of light using magnifying lenses allowed early researchers to start designing instruments to visualize small objects that had not been previously known. Some of these previously unknown objects include cells, bacteria, and many protists. Bending light is the same technology which allows for corrective lenses such as glasses and contact lenses to assist those individuals with vision problems. Early researchers started looking at a large diversity of objects, making discoveries which were only possible using these new microscopes. They were looking at pond water, the plaque from scraping their teeth, thin layers from bottle corks, even the sperm cells in semen. The way these microscopes worked was to bend rays of light that were bouncing off of a tiny object in such a way that it generated a larger image. There are still limits to how small of an object a light microscope is able to see. The limit comes from the light itself, which is why there are some very powerful microscopes that do not use light at all, but use beams of flowing electrons or nuclear magnetic resonance to visualize objects which are too small to be seen with light.

© Palo_ok, 2013. Used under license from Shutterstock, Inc

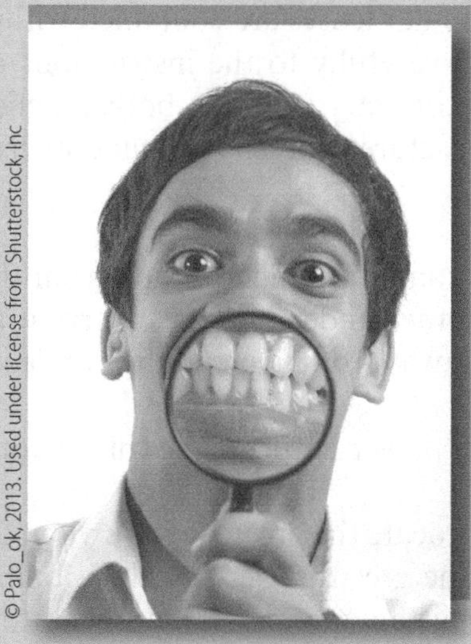
© Palo_ok, 2013. Used under license from Shutterstock, Inc

The light microscopes that we will be using in today's lab are much more powerful than the first microscopes, but there are still many objects which we have discussed in lecture that we will not be able to see. The limit of a light microscope's ability to magnify objects is at around the size of the smallest bacteria. Light microscopes cannot be used to see individual molecules of DNA, protein or water. Light microscopes are not able to see viruses or protein fibers or sugars. But we will be looking at a variety of common objects that you likely have never seen before at such high magnification.

There are a few general rules when it comes to microscopes and the reason for these rules is to prevent damage to these very expensive tools.

Common Notes on Using a Compound Microscope

- Always carry a microscope with two hands, one hand on the arm, the other under the base.
- Begin every examination at the lowest magnification, the scanning objective with the red line.
- Always return to the scanning objective (red line) before removing the slide from the stage.
- Clean lenses only with microscope lens paper, not paper towels, tissues, cloth, or kimwipes.
- Never completely unscrew anything on the microscope. The dials can be adjusted, but no screws should be removed.
- You do not need to wear glasses if they correct for near- or farsightedness.
- If you have any problems with your microscope, err on the side of safety and consult your instructor.
- Always remove the slide you have examined before storing your microscope.
- Remember to replace the dust cover and return your microscope and its power cord to their appropriate space at the end of the lab period.

PROTOCOL

Lab Activities

Activity 1: Introduction to the Compound Microscope

The compound microscope is a delicate and expensive instrument. This type of microscope projects light from below a specimen. This light is then concentrated to illuminate the specimen. In order for an image to be visualized with a compound microscope, the specimen must be transparent in certain areas, so that the light can pass through to be magnified by the lenses. The light is magnified by lenses within an objective above the specimen and then again within the ocular lenses. The ocular lenses are contained within the eyepieces you use to examine the specimen. Listen carefully to the instructions of your lab instructor. Learning to use your microscope correctly will take both practice and patience. In this activity you will learn about the structures of the compound microscope.

1. One member of the lab group should remove the compound microscope from the cabinet, gasping it with one hand on the arm, the other under the base. Carry it to your lab bench and put it down gently, with its arm away from you and the eyepieces toward you.
2. Another member of the lab group should obtain a power cord for the microscope from the appropriate drawer.
3. Plug the microscope into the outlet on your bench. Locate the green power switch on the base and turn it on. If the red button next to the green power switch is illuminated, please turn the red light off by pressing the button once. We do not want the red button to be illuminated during this lab activity.

4. The amount of light emitted from the light source is controlled by two features on the base. There is a black plastic dial which is a rheostat, which controls the amount of power going to the lamp. By rotating this dial, more or less power will be traveling to the light source. There is also an iris diaphragm on the light source itself (controlled by a ring) which allows more or less light to be seen.
5. As light leaves the light source, the light is directed through a lens into a structure called the condenser, which is located above the light source and hangs from the stage. The condenser includes a series of filters and focuses light onto the specimen. The higher the number on the condenser, the more tightly the light will be focused on the central portion of the field of view. The 'BF' setting tells the condenser to diffuse the light evenly across the entire field of view. For our lab activities today, we want to ensure that the filter is set to "BF" which stands for "bright field." If there is a number displayed on the filter such as 10, 20, 40, or 100, please rotate the filter so that BF is displayed.
6. The ring which surrounds the light source controls the iris diaphragm. Like the iris of your eye, it opens and closes to allow more or less light past. Twist from the right to left while looking at the light. Record your observations on the lab worksheet.
7. The main part of the stage is a platform on which to place the slide containing your specimen. The slide is held in place by a claw-like stage clip located on the left-hand side of the platform. Open and slowly release the clip. When inserted, the slide should be positioned all the way back against the bracket of the slide holder.
8. The rear portion of the stage is movable rather than stationary, so it is called a mechanical stage. This allows you to position the specimen directly under the objective, using two stage knobs located just below the stage. If you imagine the platform of the stage as a piece of graph paper, one knob moves the mechanical stage in the *x* plane (right or left) while the other moves it in the *y* plane (away from or towards you). Rotate these knobs to get a feel for how they function. Record your observations on the lab worksheet.
9. Above the stage and located on the head of the microscope is a revolving nosepiece. The objective lenses (objectives) project out and down from this nosepiece. These are the first structures that magnify the specimen according to the magnifying power printed on the outside of each lens. Your microscope contains five objectives. Grasp the knurled ring surrounding the nosepiece and rotate it. Notice how each objective "clicks" into position.
10. On the lab worksheet, the table in question 3 lists the name and function of the five different objectives on these compound microscopes. The shortest objective (the one with the red line on it) is the scanning objective; next is the low-power (with a yellow line) and then the high-power objective (with a green line). This is the highest objective that we will be using in this lab. We will not be using the second highest magnification objective (blue line) or the oil-immersion objective (white line) because this objective requires the use of microscope oil on the surface of the slide in order to focus the specimen correctly. Record the magnification of all of the objectives in the table on the lab worksheet.
11. On top of the microscope head are the binocular eyepieces, or ocular lenses. Because your microscope has two such lenses, it is referred to as a binocular microscope. You can adjust these ocular lenses to position each one so that you can see more comfortably. Make this adjustment by grasping the base of each eyepiece with the thumb and forefinger. Push the oculars as close together as possible while looking through the eyepieces. Slowly move the oculars away from each other until you see one circle of light.
12. The ocular lenses also provide another means of magnification. Ocular magnification is printed on the outside of the ocular lenses (the number followed by an X). Record this number on the (Question 3) table in the lab worksheet. The total

magnification is the magnification of the objective lens in place, multiplied by the magnification provided by the ocular lenses. Calculate the total magnification power of each objective and record the numbers in the table on the lab handout.

13. You can focus the microscope by using the coarse- and fine-adjustment knobs located on either side of the arm near the base. The larger of the two knobs provides coarse adjustment, used when the scanning or low-power objective is in place. Always begin focusing with the scanning objective in place, then adjust your focus as you progress to the low- and high-power objectives. Turn the coarse adjustment knob. Record your observations on the lab worksheet. Never use the coarse adjustment when the high power objective is in place!

Activity 2: Observing and Drawing objects under magnification

1. Obtain a slide with object(s) which you wish to observe, then bring that one slide to your microscope.
2. Make sure the scanning objective (red line) is in place and set the slide on the stage, positioning it into place with the stage clip.
3. Using the stage knobs, make sure that the object to be viewed is centered under the light emitted from the condenser.
4. Looking through the eyepieces, use the coarse and fine focus knobs to bring the image into focus. (Note: Be patient, this may take some practice. It usually involves turning the coarse adjustment knob so that the stage moves closer to the scanning objective.) Unless you are looking at a slide of dust, your sample should not look like dust particles. This means you are probably looking at the upper surface of the slide. You need to adjust the lens to be even closer to the specimen to focus on the sample rather than the upper surface of the slide.
5. Once the image is in focus with the scanning objective, then switch to the next objective (yellow line) which is the low power objective. (Note: On these microscopes you do not need to lower the stage or change the focus when switching between the objectives.)
6. It should only require movement of the fine adjustment knob to bring the image into focus under the low power objective.
7. You may choose to switch to the high power objective to observe the item under even higher magnification. (Note: Be sure not to move the coarse adjustment knob while the high power objectives are in place. Also, we will not be using the two highest power objectives (blue and white lines) in this lab activity.)
8. Draw a picture of the item under your preferred magnification (scanning, low power, or high power) and be sure to record the total magnification of the item you are drawing.
9. When everyone in the lab group has drawn the item, return the microscope to the scanning objective (red line), remove the slide, and exchange that slide for another slide.
10. In the course of this assignment you are to draw images of objects from five different slides.
11. For the cleanup for this lab, return the slides that were observed to their boxes, return the plastic dust cover onto the microscope, place the microscope into a shelf at the side of the room, and return the power cord to the drawer that you got it from.

Name ______________________________ Date ______________

Microscope Usage—Student Worksheet

QUESTIONS

1. Twist the ring controlling the iris diaphragm on the light source from the right to left; what happens?

2. Observe how the stage knobs affect the mechanical stage. How will they help you examine your specimen?

3.

Name of Objective	Magnification	Function	Ocular Magnification	Total Magnification
Scanning (Red)		To locate the specimen		
Low-Power (Yellow)		To examine the specimen in some detail		
High-Power (Green)		To examine the specimen in more detail		
Second Highest Power (Blue)		To examine the specimen in even greater detail		
Oil Immersion* High-Power (White)		Highest magnification for a compound microscope		

*Oil Immersion lenses will not be used in today's lab activity but the table must still be completed.

4. Turn the coarse-adjustment knob. What structure of the microscope is moving?

Name ______________________ Date ______________

5. Label the diagram.

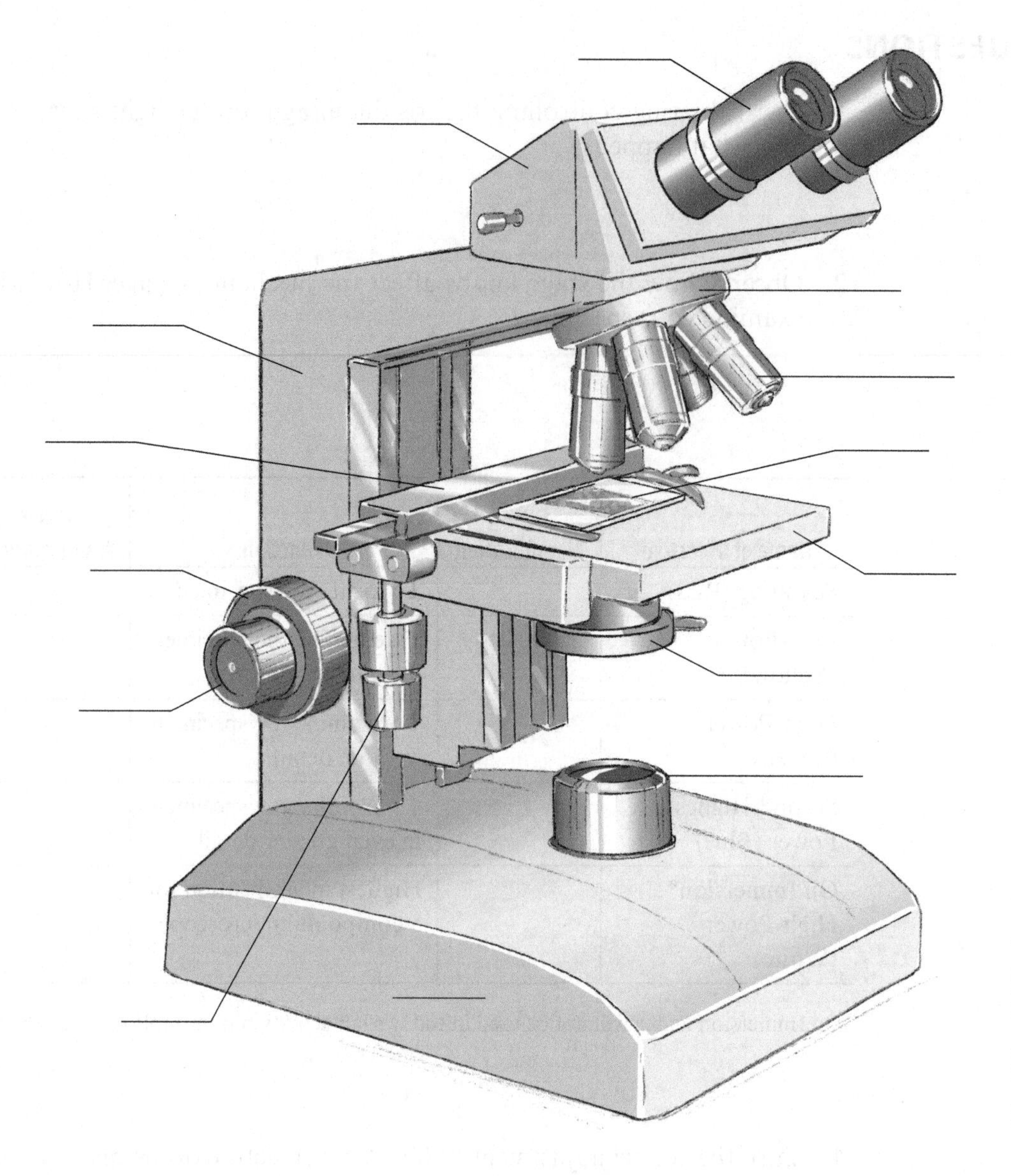

© Kendall Hunt Publishing Company

6. What two structures contain lenses that magnify a specimen in the compound microscope?

7. Two structures control the amount of light emitted by the compound microscope. Name these two structures and describe their function.

Name ______________________________ Date ______________

8. Name the structure responsible for determining how tightly light is focused on onto the specimen.

9. Write the formula for calculating total magnification.

10. Which objective should be in place when you're initially observing a specimen? Why is it important to remember this when you're focusing the microscope?

11. Draw five different objects that you observed under the microscope. Identify the object and be sure to record the total magnification that you used for each.

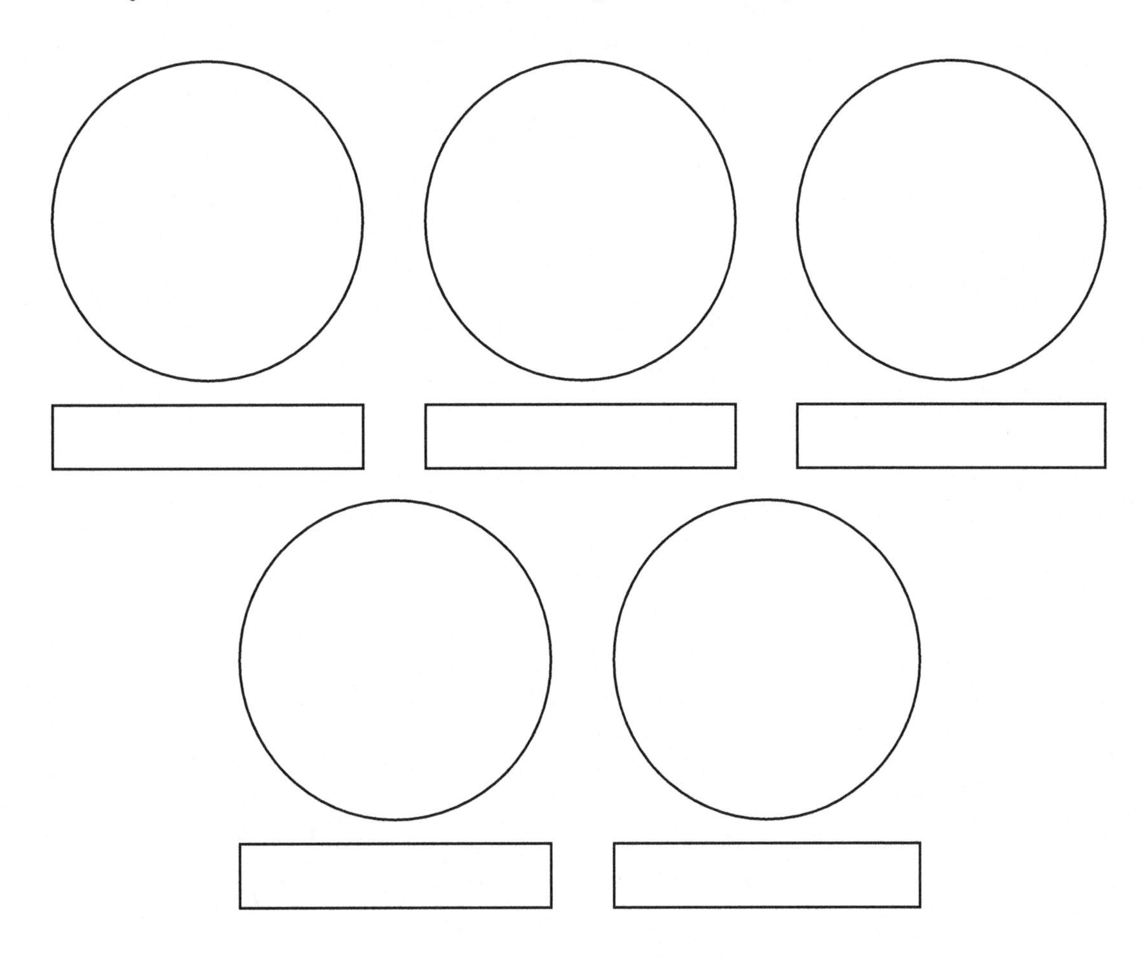

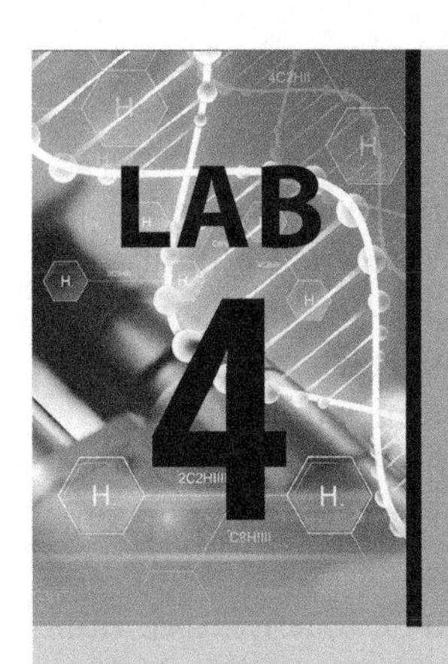

LAB 4 Osmosis and Diffusion

BACKGROUND

When we were discussing measurements earlier, we talked about a few measurements which were ratios between two different values. An example of a measurement which is a ratio is density, mass/volume. Another example of a measurement unit which is a ratio is velocity, which is the distance that an object travelled given a specific amount of time. In today's lab we will be discussing another ratio measurement known as concentration. This measurement is specifically used when we are talking about solutions. A solution is a combination of at least two substances, the material being dissolved, called the solute, and the material doing the dissolving, which is known as the solvent or liquid portion of the solution. Any solution in which water is the solvent is called an aqueous solution.

The concentration of a solution is a ratio between the amount of solute in that solution and the amount of solvent. If you had a solution that had one gram of salt dissolved into 100 ml of water and another solution that had five grams of salt dissolved into 500 ml of water, we would say that these two solutions have the same concentration, because even though you have five times as much salt in the second solution, you also have five times as much water, so the ratio of salt to water is the same in the two solutions. If you have one gram of salt dissolved into 100 ml of water and another solution that had five grams of salt in 200 ml of water, they would have different concentrations because the ratio of salt to water would not be the same for both. If you were to place these two solutions into the same container so that they were in contact with each other, we would have a concentration gradient and something very interesting would happen. The components of the solutions would have a net movement resulting in an even mixing of the two solutions. Simply due to the movement of molecules in a liquid or gas solution, anything dissolved into those solutions will always have a net movement from high concentration toward low concentration. What that means is that solutions of differing concentrations in contact with each other will always try to equalize their concentrations, achieving a state called equilibrium. This equilibrium is achieved by substances moving from a high concentration to a low concentration. Both components of the solution are moving in this process, both the solute and the solvent. Both of these components will move from where they are in a high concentration to

© Mariyana Misaleva, 2013. Used under license from Shutterstock, Inc.

where they are in a low concentration. This described movement, from high concentration to low concentration, is known as diffusion. Diffusion is a property of all liquid and gas solutions simply due to the movement of their molecules. Even when equilibrium is reached, the molecules of liquids and gasses do not stop moving; the molecules are still moving, but in no specific direction. You get just as many molecules moving in one direction as you get moving in the opposite. This is known as dynamic equilibrium.

Since concentration is a ratio between solute and solvent, a solution with a high solute concentration has a low solvent concentration and vice versa. There are specific terms used to describe solutions that differ in their concentrations when compared with other solutions. If you have a solution that has a higher solute concentration than another solution, we would say that that solution is hypertonic. Since concentration is a ratio, a hypertonic solution will have a lower solvent concentration than the compared solution. The solution that has a lower solute concentration than another solution is called a hypotonic solution. That hypotonic solution will have a higher solvent concentration than the compared solution. When two solutions have the same concentration, we say that they are isotonic. When solutions have the same concentration (isotonic) then equilibrium has been achieved. We can only use these terms (hypertonic, hypotonic, and isotonic) when we are comparing two or more solutions. This term cannot be used to categorize a single solution.

Sometimes you will have solutions that will be in contact with each other, but separated by a barrier or membrane that would allow certain molecules to pass through while blocking others. This is a common situation in cells where the plasma membrane separates the inner aqueous solution of the cell (cytoplasm) from the aqueous solution outside of the cell. The plasma membrane is permeable to some substances, but impermeable to others. It is for this reason that we say it has selective permeability. Water is one of the substances that can pass through the plasma membrane, but electrically-charged ions and larger sugars and amino acids are not able to pass through the membrane. If the concentration of ions is higher in one of two solutions separated by a selectively permeable membrane the diffusion of the ions may blocked by the barrier, but the diffusion of the solvent (water, in this case) would not be blocked. The diffusion of the solvent of a solution is called osmosis. Diffusion always is a movement from high concentration toward low concentration, and osmosis is no exception. The key to

© ankomando, 2013. Used under license from Shutterstock, Inc.

remember, though, is that when you have a high solute concentration, that solution has a low solvent concentration. Because of this, water will always flow from the hypotonic solution toward the hypertonic one, even though the solute would move in the opposite direction if it could. Animal cells change in size when the concentration of the cytoplasm does not match that of the environment. Since water always flows toward the hypertonic solution, if an animal cell is placed into a hypertonic solution, it will shrivel because the water will flow out of it. If an animal cell is placed in a hypotonic solution, it would swell because the water would rush into it. So much water could try to enter the cell that the cell could burst. If an animal cell is placed into an isotonic solution, it will not change its shape because just as much water would rush into it as out of it. The reason that most IV's are of a saline solution is because they want the concentration to match that of normal human cytoplasm, so that it will not put further strain on the patient's cells.

In today's lab we will be using dialysis tubing as a selectively permeable barrier to test osmosis. Dialysis tubing is filled with microscopic holes that are small enough to allow water through the walls of the tubing, but not sucrose molecules. We will be filling dialysis tubing bags with solutions of sucrose and water in various concentrations, then getting an initial mass measurement of the filled bags. We will then set a timed session of osmosis, by placing those filled dialysis tube bags into water for 30 minutes. At the end of 30 minutes we will take a second mass measurement of each of the bags and compare the masses. The results of our experiments will show the direction in which the water flowed and the rate or speed at which the water travelled as well.

PROTOCOL

Activity: Measuring rates of Osmosis

Materials needed by each lab group

1 Plastic cup with 6 strips of Dialysis tubing
12 Plastic cups (2 labeled "Water," 2 labeled "0.2 M," 2 labeled "0.4 M," 2 labeled "0.6 M," 2 labeled "0.8 M," 2 labeled "1.0 M")
1 25 mL graduated cylinder

Shared Lab Materials

6 Sugar-Water solutions of varying concentrations
Scales

1. Upon obtaining the materials for this lab, fill one of each of the labeled cups ¾ of the way full with water. (Note: This will be 6 of the 12 cups, one of each label, "Water"-"1.0 M".)
2. Take one of the strips of dialysis tubing and tie a knot in one end. (Note: The dialysis tubing looks like a strip of plastic, but is it actually a tube that can be opened by rubbing the end of the strip between your fingers. The dialysis tubing itself is covered in tiny holes which allow water to pass through, but not sugar molecules. Lotions, oils, and dirt from the hands could clog these tiny holes, so please wash your hands before handling the dialysis tubing.)
3. Open the other end of the dialysis tubing by rubbing your fingers together with the tubing between them. You now have an open pouch due to the knot on the other end of the tube. (Note: If the tubing is starting to dry out, it can be very difficult to open. For that reason it is best to leave the dialysis tubing in the water until it is

being used. If you are having difficulty opening the dialysis tubing, return it to the water briefly to moisten it.)

4. Measure 25 mLs of water using the graduated cylinder and pour the water into the dialysis tubing.
5. Knot the other end of the dialysis tube in order to make a sealed pouch containing the water. (Note: Make sure that the second knot is not too close to the solution, to allow for a possible increase in volume of the internal fluid during the experiment.)
6. Blot the sealed dialysis tube bag dry and get a mass measurement of the bag using a scale at the side of the room. Record the mass (in g) of the filled tubing under "Initial Mass Measurement" for the solution labeled "Water" in the Data Table.
7. Place the tube into the empty cup labeled "Water," then repeat steps 2–7 for the other five solutions, placing each filled bag into the properly labeled empty cup. (Note: One of each cup should be ¾ full of water and one empty, we will collecting all of the tubes in the labeled empty cups until all lab groups are completed with the first part of this experiment. When filling the other dialysis tube bags with their sugar solution, start with the lowest concentration "0.2 M" and work up to the highest concentration "1.0 M" so that you do not need to wash your graduated cylinder between solutions.)
8. Once all lab groups have recorded their "Initial Mass Measurement" for each of the 6 solutions, we will all transfer the tubes into the water-filled labeled cups at the same time. (Note: Even though each set of cups has a different label, they are all filled with the same water. The only thing that is different is the solution within the dialysis tubing.)
9. Everyone will let the dialysis tube bags sit in the water for 30 minutes. (Note: This 30 minute incubation period is synchronized between all of the groups because we waited to start until each group was ready.)
10. At the end of 30 minutes, remove each tube from the water-filled labeled cup and return it to the empty labeled cup.
11. Blot dry each tube with paper towels and get a mass measurement using the same scale that you used for your first measurements. Record the new mass (in g) under "Final Mass Measurement" in the Data Table.
12. Calculate the "Change in Mass" for each solution by subtracting the Initial Mass from the Final Mass. (Note: The equation for this calculation should be as follows : Final Mass − Initial Mass = Change in Mass.)
13. Calculate the "% Change in Mass" for each solution by dividing "Change in Mass" by "Initial Mass" and multiplying by 100. (Note: The equation for this calculation should be (Change in Mass/Initial Mass), then × 100 = % Change in Mass.)
14. Record your group's "% Change in Mass" data on the board. Once all groups have entered their data, we will calculate the class' average "% Change in Mass" for each of the six solutions. You will then graph your groups data and the class average data for "% change in mass" for the six solutions.
15. Clean up for this lab involves pouring the water from the plastic cups into the sinks, then stacking the plastic cups and returning them to where you got them. All of the dialysis tubing bags can be discarded into the garbage after recording their "Final Mass Measurement." Because the sugar solutions can be sticky, please wipe down your work area and the scales that were used for measurement with moist paper towels.

Name ______________________ Date ______________

Osmosis and Diffusion—Student Worksheet

QUESTIONS

1. Graph the % change in mass for your group's data and for the class average data.

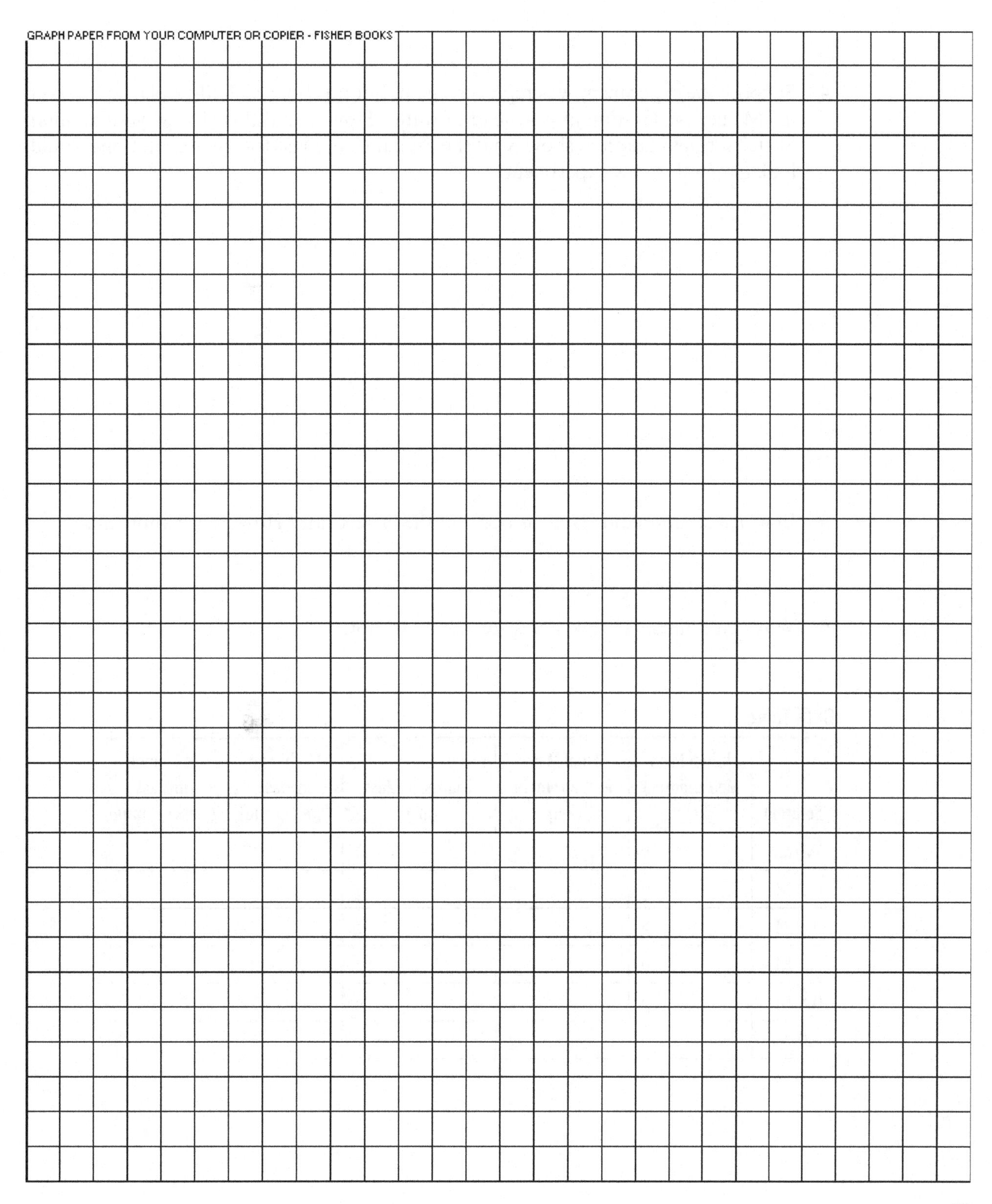

Name ______________________ Date ______________

2. What caused the change in mass of the dialysis tube bags? Was there more or less water in the dialysis tube bags at the conclusion of the experiment? Explain.

3. Was the water in the plastic cups hypertonic or hypotonic in relation to the sugar solutions found in the last five dialysis tube bags? Why do you say that?

4. Suppose the experiment was repeated again, but this time each filled plastic cup had 0.6 M sucrose solution in it instead of water. How would this change your results? Sketch a rough graph to show what the % change of mass for the six solutions would look like in this new experiment.

5. In which solution did Osmosis occur at the fastest rate? How do we know this?

6. Why can't humans drink salt water for hydration?

Data Table

Solution	Initial Mass Measurement (in g)	Final Mass Measurement (in g)	Change in Mass (in g)	% Change in Mass (Your Group)	% Change in Mass (Class Average)
Water	g	g	g	%	%
0.2 M	g	g	g	%	%
0.4 M	g	g	g	%	%
0.6 M	g	g	g	%	%
0.8 M	g	g	g	%	%
1.0 M	g	g	g	%	%

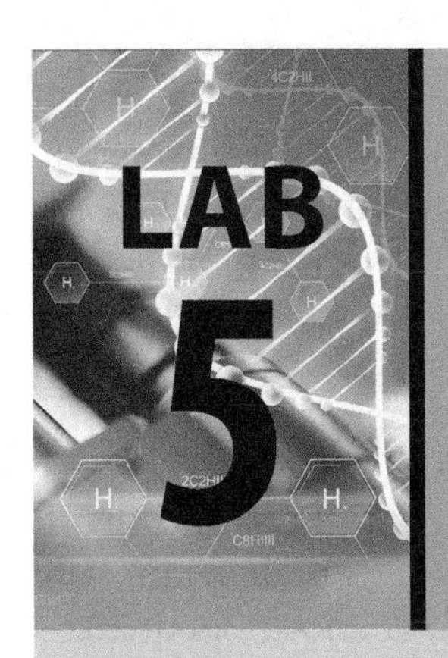

LAB 5 Enzymes in Digestion

BACKGROUND

Catalysts are substances that are able to speed up chemical reactions by reducing the amount of activation energy required to initiate the conversion from reactants to products (this is what happens in a chemical reaction). Catalysts come in many forms, some are biological in origin and others are not. We call biological catalysts "enzymes." There are tens of thousands of different enzymes in each one of your cells. The reason that so many different enzymes are needed in each cell is that each enzyme can catalyze (speed up) one specific type of reaction. The enzyme can be used over and over again as it is not consumed by the reaction, but it can only catalyze one specific type of reaction. Since we have millions of chemical reactions occurring every minute in each one of our cells, we need to have thousands and thousands of different enzymes. What causes each enzyme to be specific to its reaction is its shape. Each different type of enzyme has a unique three dimensional shape and that shape determines its function.

The shape of the enzyme determines which substrates it can interact with, much like there is usually only one shape of key that will operate a specific lock. Substrates are usually the reactants of the chemical reaction which is being catalyzed by the enzyme. Because the shape of the enzyme is so important for its function, if something were to happen to the shape of the enzyme its function would be affected as well. Most of the enzymes within a cell are proteins, so when we are dealing with enzyme shape, we are dealing with protein structure. There are four levels to protein structure. The first level is known as the primary structure. Primary structure is simply the order in which the amino acids are connected in the polypeptide chain. Recall that amino acids are the monomer building blocks for proteins. While it may not initially seem all that important, it is the primary structure that ends up determining all of the other levels of structure. The second level of structure is, appropriately, called the secondary structure. These structures are due to the amino acids forming hydrogen bonds with each

© Einar Muoni, 2013. Used under license from Shutterstock, Inc.

other which results in a repeating spiral structure known as an α (alpha) helix, or a folded sheet-like structure known as a β (beta) sheet. The third level of structure is known as tertiary structure and it is the result of the interactions of the secondary structures and the unstructured regions of the polypeptide chains known as loops. These interactions are due to hydrophobic and hydrophilic interactions between these structures. Tertiary structure is the final level of structure for a single polypeptide chain. Only if a protein is composed of more than one polypeptide chain will it have the fourth level of structure, which is known as quaternary structure. Quaternary structure is the way in which multiple polypeptide strands interact to form a single functional protein with a specific three-dimensional shape. Each of these additional levels of structure are determined by the primary structure (order of amino acids in the polypeptide chain). Another factor which can have an impact on the shape of a protein is the cellular environment that it finds itself in. Changes in temperature, pH, and ion concentration can end up having an impact on the shape (and therefore the function) of a protein. Certain enzymes in our body are only functional in particular locations because that is where the environment is compatible with their functional shape (examples are the stomach and small intestine, which have different conditions and different functional enzymes).

Chemical digestion within a living organism involves polymers being broken down into their monomer units. This specific chemical reaction is known as hydrolysis. Enzymes which catalyze this reaction are called hydrolases and are essential for digestion. There are different hydrolases that break down the different types of biological molecules. We will be observing the action of a hydrolase that specializes in breaking starches down into monosaccharides in today's lab.

Materials for each group

3 transfer pipettes
1 small multi-welled tray (for performing your study)
1 large multi-welled tray (to hold your solutions as you return to your work space)

Shared materials (everyone on your table may need to share the same bottles of these)

Starch solution/suspension (clear)
Starch indicator or detector solution (brown)
Enzyme solution
Water
Ice
Acid
Base

Note: You will be combining water, enzyme, and starch in different quantities and for different times to explore the action of enzymes. You should consistently use the following procedure:

1. Add the number of drops of water to all the wells listed in a table.
2. Add the number of drops of enzyme indicated in the table.
3. Add the starch solution and immediately start the timer.
4. Once the time has elapsed, quickly add the brown starch indicator. It will stop the reaction. Your results will be stable over time, so you don't have to immediately observe the results.

Name ______________________________ Date ______________

Enzymes in Digestion—Student Worksheet

QUESTIONS

Activity 1: How do the solutions in this study behave?

In order to make reasonable comparisons, we always use water to equalize the volume of liquid in each well. Fill in the first four wells of your multi-welled plate as detailed in Table 1, and complete the table with your color result. The numbers under the 'Water,' 'Enzyme,' 'Starch Solution,' and 'Starch Indicator Solution' columns reveal to you the number of drops to add to each well. Work quickly and as a group to ensure that a minimal amount of time passes between combining the enzyme and starch solutions and putting in the starch indicator. Swish your solution around and view it over a white background to consistently assess the resulting color. Use the information in the table and your results to answer the questions below the table.

Table 1—Number of drops to add to each well for Activity 1, along with the time before adding starch indicator solution. Enter results in last column

Well Number	Water	Enzyme	Starch Solution	Time before Adding Indicator (minutes)	Starch Indicator Solution	Color Result (clear, yellow, brown, black, etc.)
1	0	4	6	0	1	
2	6	4	0	0	1	
3	4	0	6	0	1	
4	10	0	0	0	1	

1. What color does the solution turn when starch is present?

2. What color does the solution turn when starch is absent?

3. The enzyme should not have affected the color of the solution because we immediately stopped the reaction by adding starch indicator solution. If we had let the enzyme work in the wells for a minute and then added our indicator, which well would show a different result from what you observed? Why?

Activity 2: What is the effect of the enzyme on starch over time?

Now we can confirm your hypothesis in question 3 by allowing the enzyme to work over time. We need to have a control with no enzyme—just starch—so that we know that we are seeing enzyme digestion and not just the starch breaking down over time naturally. Fill in eight wells of your multi-welled plate as detailed in Table 2, carefully timing your starch indicator addition from the time when you mix starch and enzyme together. You can do more than one well at a time, as long as you can accurately deliver the starch indicator at the right time.

Name ______________________ Date ______________

Table 2—Number of drops to add to each well for Activity 2, along with the time before adding starch indicator solution. Enter results in last column

Well Number	Water	Enzyme	Starch Solution	Time before Adding Indicator (minutes)	Starch Indicator Solution	Color Result (clear, yellow, brown, black, etc.)
5	0	4	6	0.5 (30 s)	1	
6	4	0	6	0.5 (30 s)	1	
7	0	4	6	1	1	
8	4	0	6	1	1	
9	0	4	6	1.5 (90 s)	1	
10	4	0	6	1.5 (90 s)	1	
11	0	4	6	2	1	
12	4	0	6	2	1	

4. Did you observe any color changes over time in the wells that contained no enzyme (well numbers 6, 8, 10, and 12)? Was this result expected? Why or why not?

5. Which well did you expect to have the least starch left when you added the indicator, and why?

6. Look at your results in Table 2. Did you observe what you expected to see given your answer in question 5?

Activity 3: How does temperature affect enzyme function?

Recall that temperature dictates the average speed at which atoms or molecules are moving. For this activity, you will use a chilled starch and enzyme solution and compare enzyme activity with that in room temperature solutions. You may need to clean out your multi-welled plate to start fresh if you are out of wells.

Asterisk ("*") in Table 3 means that you should use the cold samples located on one of the side tables in the room. No asterisk means to use the room temperature samples at your table. Be very careful that you don't have your hands on the cold samples in your pipette or well plate; the volume is so small that you will quickly warm them up. If you want to be doubly sure that your cold wells are cold, you can try to balance the cold wells on a piece of ice during the minute the experiment runs. It is fine to do the cold wells separately from the room temperature wells, since you will be able to compare the colors due to the starch indicator solution stopping the reaction.

7. What was the purpose of wells 2 and 4: those which contained starch but no enzyme?

Name ______________________ Date ______________

Table 3—Number of drops to add to each well for Activity 3, along with the time before adding starch indicator solution. Enter results in last column. * means to use the cold samples, making sure to keep them cold

Well Number	Water	Enzyme	Starch Solution	Time before Adding Indicator (minutes)	Starch Indicator Solution	Color Result (clear, yellow, brown, black, etc.)
1	0	4*	6*	1	1	
2	4*	0	6*	1	1	
3	0	4	6	1	1	
4	4	0	6	1	1	

8. Which well would you predict would digest starch the fastest, and why?

9. Briefly summarize your conclusion from this activity in complete sentences. Make sure that you do not simply recapitulate the results, but draw a conclusion about what the enzymes are doing in warm versus cold solutions. Imagine the enzymes as workers who adjust the speed of their movement in response to the temperature.

Activity 4: How does pH affect enzyme function?

Protein shape is sensitive to pH. Since enzymes are proteins, we can look at how pH changes enzyme activity by making the solution more acidic or basic. When you interpret your results, remember the yellow color that the solution takes on when you have very little starch. If you get a color different from any previous result in this lab, it may be that our indicator may not work well at certain pH levels.

Table 4—Number of drops to add to each well for Activity 4, along with the time before adding starch indicator solution. Enter results in last column

Well Number	Water	Acid (HCl)	Base (NaOH)	Enzyme	Starch Solution	Time before Adding Indicator (minutes)	Starch Indicator Solution	Color Result (clear, yellow, brown, black, etc.)
5	2	0	0	3	6	1	1	
6	1	1	0	3	6	1	1	
7	0	2	0	3	6	1	1	
8	1	0	1	3	6	1	1	
9	0	0	2	3	6	1	1	

Name ______________________________ Date ______________

10. Which well represents a neutral pH (pH = 7)?

11. Did the enzyme seem to work best at neutral pH, more acidic pH, or more basic pH? You will be able to see the pH of the well by looking at the number of acid or base solution drops added. For example, the well with two drops of base and no drops of acid would be a strong base. If there were no differences, you may need to repeat the experiment and add the starch indicator solution in 30 seconds. The lab temperature can vary across semesters.

12. We would expect an enzyme to work best at the pH of the place where it is located in the body. The pH of some locations within the body are: saliva at pH = 7, lysosomes at pH = 4.8, and the cytosol at pH = 7.2. Based on your results and the information above, where would you propose these hydrolase enzymes come from in the body, and why?

13. A scientist is doing enzyme assays on galactase, which uses galactose as its substrate. He or she chose a variety of substrate (galactose) concentrations in solutions that contain a common enzyme (galactase) concentration and then observed the resulting reaction rate.

Table 1. Raw data from the experiment

Substrate concentration	0%	10%	20%	40%	60%
Enzyme concentration	1%	1%	1%	1%	1%
Reaction rate (g/s)	0	7	10	12	12

a. What is the independent variable?

b. What is the dependent variable?

c. What is the constant in this study?

d. Draw a graph showing the galactose concentration vs. the reaction rate in Table 1. Be sure to label your axes and provide a figure caption.

Name ______________________________ Date ________________

e. How is the reaction rate related to the enzyme concentration, if you can tell?

f. How is the reaction rate related to substrate concentration, if you can tell?

g. What is happening at the enzyme level to produce the results from the experiment? HINT: Think about what is physically happening within the solution, with the enzymes as workers and the substrate as work to do.

LAB 6 Plant Stomata

INTRODUCTION

Stomata are little openings or pores in plants that allow the plant to exchange gasses, such as carbon dioxide, oxygen, and water vapor, with the outside environment. In order to stay alive, the plant must have oxygen to perform cellular respiration. Most plants also require large amounts of carbon dioxide to make glucose, which they in turn use as the input for cellular respiration and as a building material. One building material they make with glucose is cellulose, where the plant strings hundreds of glucose molecules together to make a polymer. Cellulose is the main component of cell walls, wood, and paper.

Stomata are made up of two special cells that are part of the epidermal cell layer. Epidermal cells cover and protect the plant from dirt, microbes, and desiccation. The two special cells flanking the stomata are called 'guard cells' because they behave like a guard at a gate, choosing whether to open and allow gases through or close the gate and deny entry or exit. An individual leaf can have hundreds of stomata on its surface, and each set of guard cells individually makes the decision to be open or closed based on the immediate situation.

The decision that guard cells make to allow substances in or out is critical to the plant's survival. As mentioned earlier, plants must have a source of carbon

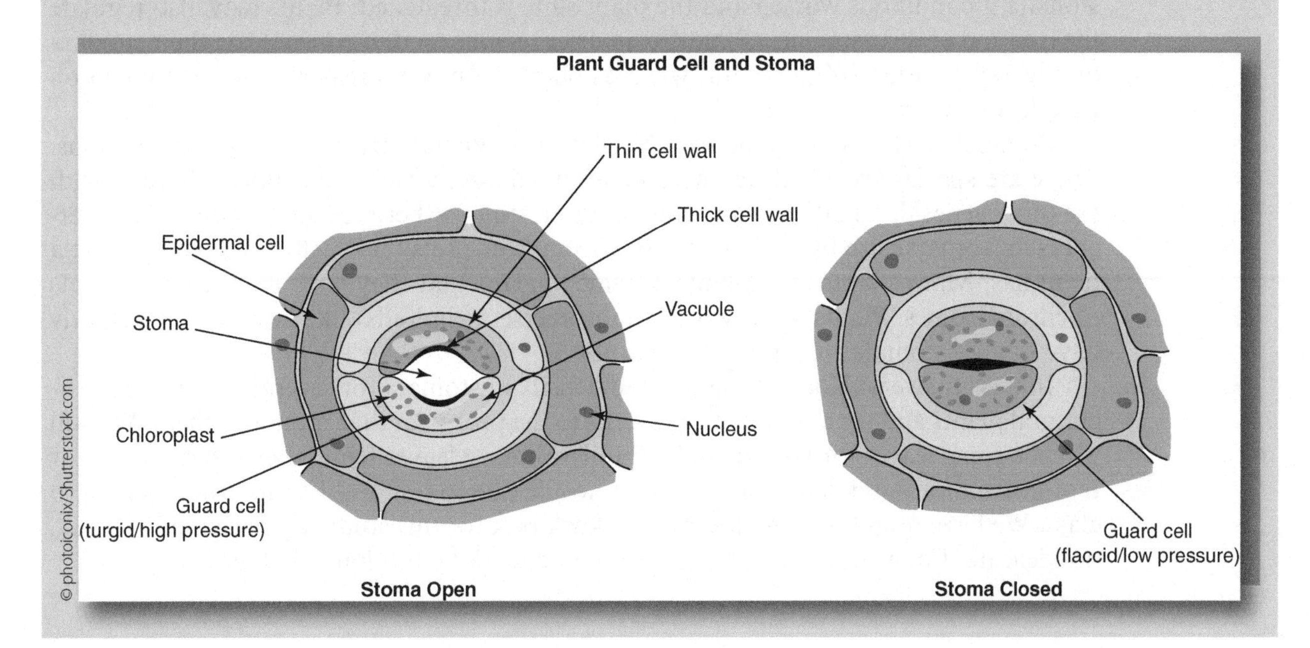

© photoiconix/Shutterstock.com

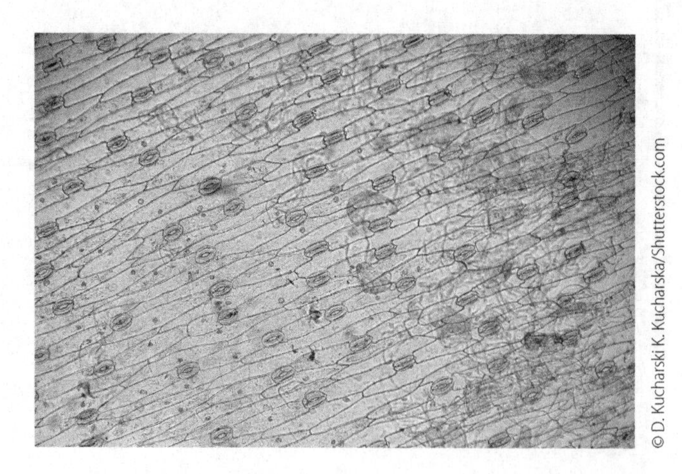

© D. Kucharski K. Kucharska/Shutterstock.com

dioxide to make sugar, so if the stomata are always closed, the plant will starve for lack of glucose and ATP. However, there is a trade-off. When stomata are open, they allow water vapor to escape from the leaf. If the plant's access to water is limited, the water loss through the stomata must be carefully balanced against the benefit of obtaining carbon dioxide. So guard cells will tend to be open when the plant has plenty of water and the carbon dioxide levels are lower in the leaf. They will tend to close when the plant is:

- stressed by drought
- surrounded by very hot and dry air
- in bright sunlight
- experiencing high carbon dioxide levels inside the leaf at that location (suggesting there is no need for additional carbon dioxide at that time).

The plant can also control how much gas (including water) will be exchanged through the stomata by regulating the number of stomata and where they are located. For example, a plant that lives in a wet environment may have a high density of stomata (that is, more stomata for each unit area of leaves) than a plant that lives in a dry environment. In the desert, there are even plants that not only limit the number of stomata on each leaf, but they can make a sticky resin to permanently block some of the existing stomata if conditions worsen and the plant's life is threatened. Plants may also regulate the location of their stomata. Plants from drier regions tend to concentrate their stomata on the lower surface of the plant, whereas plants from wet areas may have stomata on both leaf surfaces.

Stomata may also vary in size. Variety in stomatal size has many explanations. There are species-specific differences in stomatal size, which are simply inherited with the other genes that make a plant unique. Interestingly, there are also predictable differences in stomatal size by the number of chromosome sets (also called the ploidy) that a plant has. Many plants are polyploid, which means that they have more than two sets of chromosomes. Plants with two chromosome sets (diploids) like humans consistently have smaller stomata than polyploid plants.

In this lab, we will look at the size and density of stomata for several desert and horticultural plants and try to link these patterns to the processes that produced them. We will provide some leaves for you to study, but if you have leaves from plants that you eat or that live in or around your home that you would like to test, feel free to bring them in to class. We have found that the leaves that work best for this study are reasonably sturdy, not delicate. For example, a cabbage leaf would work better than a lettuce leaf.

Materials (per group)

1 bottle clear nail polish
1 sample (as available) of each leaf type
1 paper towel
1 ruler
1 pair scissors
Clear packing tape (must be completely translucent)
3 microscope slides
1 fine Sharpie

Methods

1. Make an impression of the stomata: Obtain a sample of each leaf type that is available, keeping track of which is which by writing on the paper towel. If you have your own sample, include it. You will use clear nail polish to paint a thin and continuous layer on the lower side of each leaf. The lower side of the leaf is usually less shiny, less green, and the veins tend to protrude more than on the top. Carefully paint each leaf and set it on the paper towel with the sticky side up to dry. Please close the nail polish tightly after you are finished. Do step 2 while you wait for the polish to dry.
2. Create your hypotheses: Using a digital device, look up the plants that you are studying to start filling out Table 1. Search the name of the plant and the word 'range' or 'habitat' to find out where these plants typically live (house plant or gardens or fields are possible habitats). Often a search for the plant with the term 'ploidy' will uncover the number of chromosomes. Often a diploid will be shown as '2n.' If a plant has four sets of chromosomes, it might be listed as '4n,' etc. After you find the information, fill in the blank cells in the table with your predictions based on what you know about how habitat can affect stomatal density and how ploidy can affect the size of stomata.
3. Make the slides: Cut a piece of tape approximately 1 cm to 2 cm in length for each sample. Do not touch the middle of the tape; only the very edges should have your fingerprints on them to avoid obscuring the impression of the leaf cells. Gently rub your tape on the dried nail polish, and peel up the tape by its edge, taking the nail polish along with the tape. Attach 2–3 pieces of tape side by side to each slide, using your Sharpie to label the sample on the slide.
4. Collect your data: Obtain a microscope and bring it back to your station. As you look at each sample, write your data in Table 2. You may need to adjust the lighting in the microscope to see the samples well. You will always start with the lowest power objective lens to find a good area of the sample to work with. For each sample, you should observe the following:
   a. Stomatal density: Obtain a view of your sample using the 40x objective lens (400x total magnification), which does not contain any edges of the sample (in other words, the sample fills your entire view), if possible. Count the number of stomata that can be seen in the field. Record the number on Table 2, then find a second section of the sample and record the number of stomata in that portion of the sample. Average the values for each plant to complete the table. By keeping the magnification constant, we know that we are always sampling the same area for each, so the number of stomata will indicate density.
   b. Stomatal size: After looking at all the stomata at 400x magnification, classify each one as having small, medium, or large stomata. Have at least two members of your group write down the designations so you can revisit any for which you disagree. Record your data in Table 2.

c. Presence of stomata on upper surface of the leaf: For the leaves that had the highest density and the lowest density of stomata, paint nail polish on the upper surface of the leaf. Allow the polish to dry, use tape to remove the polish, and then assess the density of stomata at 400x magnification for the top of the plant. Record your results in Table 2. Note that you only have to do this for two plants, not all of them.

5. Answer the following questions. Return your microscope and cord to the appropriate cabinets, remove the tape from the slides, rinse them with soap and water if necessary to remove debris, and dry them to remove spots. Dispose of the paper towel and plant materials in the trash. Return the remaining items to the area where you picked them up.

Name ____________________ Date ____________

Plant Stomata—Student Worksheet

QUESTIONS

Table 1—Plants we are sampling and their habitat and ploidy. For predicted stomatal density, indicate whether you expect high, medium, or low density of stomata on the leaves of each plant based on the plant's habitat. For the predicted size, indicate whether you expect the stomata to be small, medium, or large based on the plant's ploidy and habitat

Plant	Habitat	Ploidy	Stomatal Density	Predicted Stomatal Size

Table 2—Raw data and computations for stomatal density. Average density should be calculated by averaging the two samples for each species

Plant	Stomatal Density		Average Density	Size (small, medium, large)	Density on Top of Leaf for Two Plants
	Sample 1	Sample 2			

1. Justify your hypotheses for stomatal density in Table 1.

2. Justify your hypotheses for the predicted stomatal size in Table 1.

Name ______________________ Date ______________

3. Did you see clear differences in stomatal density based on habitat? Explain the pattern by pointing out general conclusions and exceptions.

4. Was your hypothesis regarding stomatal density supported? Name one factor beyond what you considered in forming your hypothesis that might have caused the results to deviate from what you expected.

5. Draw a graph showing the relationship between the size of stomata and the density of stomata for each plant that we studied. Do not forget your axis labels and caption!

6. Was the stomatal size more closely related to habitat or to ploidy, or to neither one? Is this the result you expected to see?

7. Did you expect to see stomata on the top of the leaf for either of the samples you studied? Why or why not?

8. Summarize your results from your study of stomata on the top of leaves. Include whether you saw stomata for each of the two samples, how that may relate to the density of stomata on the lower surface of the same leaf, and how habitat may have played a role in the outcome.

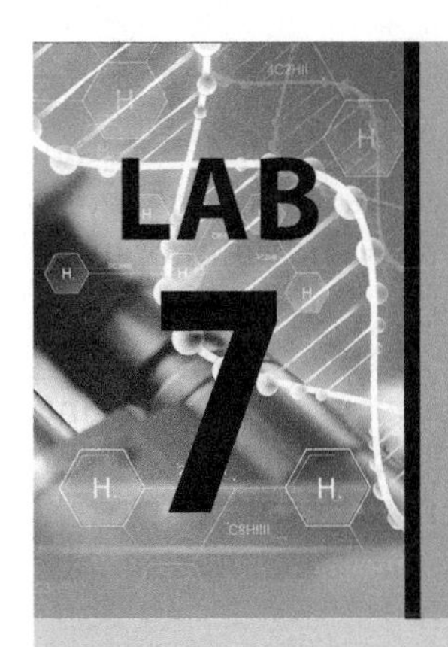

Photosynthetic Pigments

BACKGROUND

A pigment is a molecule which absorbs certain wavelengths of light (colors) and reflects other wavelengths of light. The reflected light is the light which we are able to see with our eyes and the color of the light we see tells us which light is being reflected by the pigment. The reflected light is not absorbed by the pigment. The light that is absorbed by a pigment often causes the electrons of that pigment to move to a higher energy state known as an "excited state." Those excited electrons are not usually stable at that higher energy state and will attempt to release that extra energy in the form of heat and fluorescence (emitted light of a different wavelength than that absorbed) returning them to their "ground" state. In photosynthesis, these energized electrons are harnessed to form chemical energy.

Photosynthesis is a series of chemical reactions known as a metabolic pathway. This particular metabolic pathway takes the reactants, water (H_2O) and carbon dioxide (CO_2), and converts them into the products glucose ($C_6H_{12}O_6$) and oxygen gas (O_2). Since the products are more complex than the reactants (anabolic reaction) and have more stored energy (endergonic reaction), we know that energy must be absorbed from some source other than the reactants for this metabolic pathway to proceed. The source of the energy that powers photosynthesis is sunlight and the molecules responsible for harvesting or collecting this solar energy are the pigments. The most prominent pigment involved with photosynthesis is known as chlorophyll. There are actually several pigments that are called chlorophyll and they are distinguished from each other with the addition of a letter at the end of their name (i.e., chlorophyll a, chlorophyll b, chlorophyll c, and chlorophyll d). Chlorophyll a is found in all plants and chlorophyll b is also very common. These chlorophylls appear green in color, which tells us that they absorb colors of light other than green (usually reds and blues) and that the green light is the light that is not absorbed, but is instead reflected. Green light is not used for powering photosynthesis because green light is not absorbed by chlorophyll. A plant given only green light would in essence, "starve" to death.

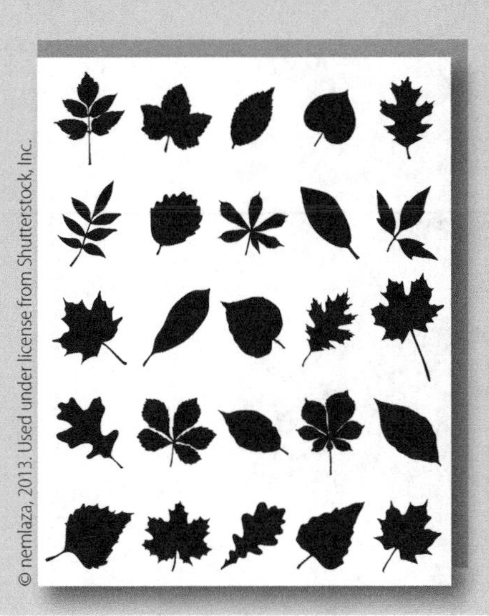
© nemlaza, 2013. Used under license from Shutterstock, Inc.

The chlorophylls are not the only pigments involved in photosynthesis, even though they are the most prominent and abundant. Carotenoids and xanthophylls are accessory pigments that are also commonly found in plants that are photosynthesizing. These pigments are often bright yellow, orange, and red in color. These pigments are responsible for the bright colors of autumn leaves and many think that these colors are shown because the tree starts making more of those pigments at this time. That is not the case. These carotenoids and xanthophylls are present in the leaves throughout the year; their presence is simply masked by the more abundant green chlorophyll. As pigments are absorbing light, they are also being broken down, so they must be constantly remade. As the tree begins to prepare to go dormant during the fall and winter, it stops producing new chlorophyll. As the chlorophyll is being broken down we start to see the accessory pigments that do not break down as quickly. So it is not that the yellows, oranges, and reds are being made at that time, it is simply that the green is not being remade so the other colors show through.

An extract from spinach leaves looks very similar to the leaves themselves, a uniform green color. But this uniform mixture actually includes several different pigments of different colors. Our goal is to separate these pigments from each other so that we might be able to see them independently. There are several ways so separate components in a mixture from each other. The method that we will be using in today's lab is known as paper chromatography. The way that paper chromatography works is through capillary action, differential solubility, and differential attraction to cellulose. Capillary action is the movement of a liquid through a dry piece of paper. The long cellulose fibers which make up paper can draw nearby liquids through the length of the paper due to the liquid's adhesion to the cellulose and cohesion to other liquid molecules. This is how paper towels work for drying up messes. The liquid that will be used for this experiment is not water, but a chromatography solution similar to nail-polish remover, or acetone. We are going to use capillary action to pull this chromatography solution through a sample of spinach pigments. As the chromatography solution moves through the pigment spot and up the filter paper, some pigments will be more soluble in the chromatography solution than others. Some of the pigments will be more strongly attracted to the cellulose of the filter paper than the chromatography solution as well. This means that as the chromatography solution moves up the filter paper, the pigments will also move up the filter paper at differing rates and the pigments will separate from each other into distinct colored bands. The pigments that travel the farthest from their starting position are those that are most soluble in the chromatography solution and the least attracted to the cellulose. It is very important to realize that in order to measure the distance that the pigments travelled, we must have a stationary starting point from which to measure

© Dulce Rubia, 2013. Used under license from Shutterstock, Inc.

the distances that these pigments moved. Ink is composed of pigments and if the reference point that we draw is made with an ink pen, the reference line may move due to the chromatography solvent. For this reason we must be sure to mark our starting line with a graphite pencil, because the graphite will not move like an ink pigment could.

The distance that the pigment moves from the starting point will be based on the distance the solvent front moves. This relationship is given a number that is called the R_f value. The R_f value is a consistent ratio between the movement of the pigment and the movement of the solvent. Each pigment will have its own R_f value and that value will not change as long as the filter paper composition and chromatography solution remain the same. It is easy to calculate the R_f value by measuring the distance the pigment travelled and dividing it by the distance the solvent travelled from the starting point. This is why we want to be sure that the mark of our starting point does not move.

Chromatography is useful any time there is a need to separate a complex solution into its component parts. When making medicines, it is important that the active ingredients are extremely pure. Chromatography is widely applied in biomedical and pharmaceutical sciences to separate substances and analyze purified compounds. In lab today, you will be doing thin-layer (paper) chromatography (TLC), which is the simplest chromatography. Other forms, such as column (flash) chromatography and gas chromatography, are more complex, but they are still based on the same principles as TLC.

PROTOCOL

Activity: Paper Chromatography of Spinach Pigments

Materials Needed by each Lab Group

1 Chromatography Strip
1 Chromatography Vial with 1 mL of Chromatography Solution
1 Spinach Leaf
1 Ridged Coin (Quarter or dime, plastic coins can be provided)
1 Ruler able to measure in cm and mm
1 Pencil
1 Pair of Scissors

1. Measure the Chromatography strip using the ruler and make a pencil line 1.5 cm from one of the ends. This will be the "starting line" from which all of the measurements in this lab activity will be taken. (Note: The chromatography strip at this point is still sensitive to oils and dirt and grease from your hands; only touch the paper strip by its edges.)

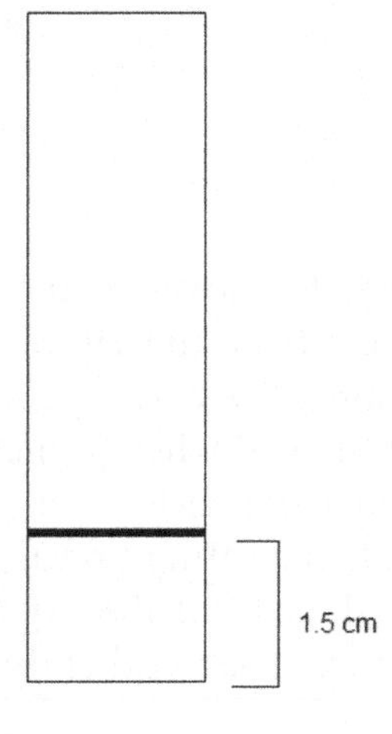

2. Using the scissors, cut the filter paper to make a pointed end on the marked side.

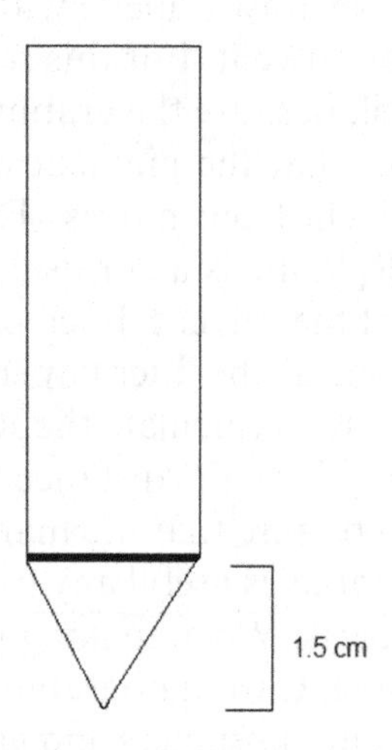

3. Using a ridged coin, crush several layers of a folded spinach leaf along the starting line, leaving a green band of pigments. This may take several rounds of crushing. (Note: We want the pigments localized around the pencil line; we do not want spots of pigment above or below the pencil line. The band of pigment will be wider than the pencil line, and that is just fine.)
4. One member of the lab group should open the cap to the chromatography vial while another member places the chromatography strip into the vial. You only want the very end of the pointed strip to touch the chromatography solvent. Quickly return the cap to the vial because the chromatography solution evaporates rapidly and is fragrant. (Note: The chromatography strip may be flexible along the pencil line due to the addition of spinach extract, make sure that the strip does not fold, because we do not want the pigments to be sitting in the Chromatography solution.)

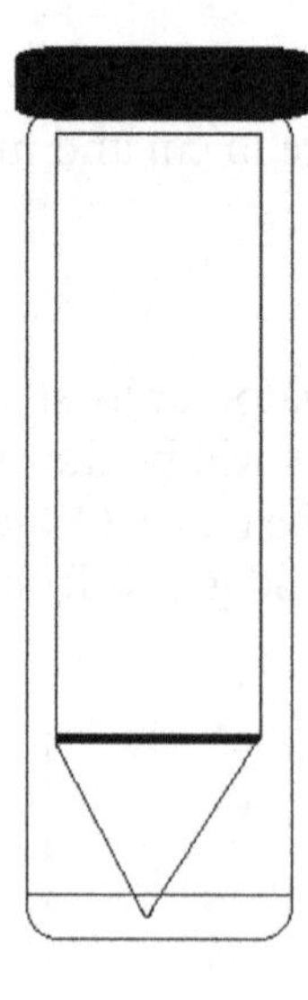

5. Allow the solvent to move up the chromatography strip, until the solvent front reaches about 1 cm from the top of the strip. It can be extremely difficult to see the solvent front against the black background of the lab benches. Place a white piece of paper behind the vial when you check the movement of the solvent front. (Note: While the solvent front moves up the chromatography strip, you will start seeing bands of pigments migrating upwards. These bands may begin migrating in strange patterns, but by the end of the experiment, there should be distinct bands of color that are different from each other due to differential migration speeds.)

6. When the solvent front reaches 1 cm from the top, the chromatography reaction should be stopped by removing the strip from the vial, recapping the vial, then quickly marking the solvent front on the chromatography strip. (Note: There are several things going on in this step so it will be best to have several lab members involved. The reason that the solvent front must be marked immediately is that the solvent will evaporate very quickly, a matter of seconds, and we need to know how far it traveled in relation to our starting line. At this point, the strip is no longer sensitive to oils and dirt, so it is alright to touch the strip while marking the solvent front's migration. The reason we need to recap the vial quickly is also due to the rapidly evaporating chromatography solution. It is essential that the reaction be stopped before the solvent front reaches the end of the chromatography strip because we need to know how far it travelled in relation to the pigments and the starting line. If the reaction is not stopped before the chromatography solution reaches the end of the strip, then the entire process must be repeated a second time, so it is in your best interest to pay attention to the solvent front and end the reaction in time.)
7. After marking the solvent front's distance, mark the center of each pigment band and then measure the distance travelled by the solvent and each of the pigments in mm in relation to the pencil line. Record this information in the Data Table.

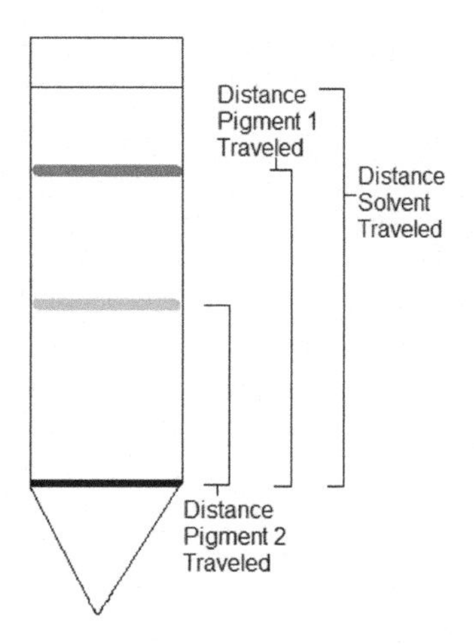

8. Calculate the R_f value for each pigment band, which is a ratio between the distance that the pigment travelled from the starting line in mm divided by the distance that the solvent travelled from the starting line in mm. (Note: the R_f value will always be less than 1, because the solvent will have travelled farther than the pigments. Sometimes you will have an R_f value of 1 if the pigment travelled along with the solvent front. The R_f value is a ratio and has no units because the mm from the two distances cancels each other out.)
9. Clean up for the lab: one lab member must staple the actual chromatography strip onto their student worksheet near the Data Table, the other members must draw a representation of their chromatography strip in the same area. The vials should not be rinsed out, but simply capped and returned to where they were picked up from. The coins, scissors, and rulers should be returned, while the spinach leaves and small paper pieces should be discarded.

Name ______________________________ Date ______________

Photosynthesis Pigments—Student Worksheet

QUESTIONS

1. How far did your solvent front travel (in mm) in relation to the starting line?

2. Which pigment travelled the farthest on your chromatography strip? Why?

3. Pigments absorb light of a certain wavelength and reflect light they cannot absorb. What does this tell us about chlorophyll and green wavelengths of light in relation to photosynthesis?

4. Why do some leaves change color in autumn?

5. What is the function of chlorophylls in photosynthesis?

6. What does the R_f value represent? If you were to perform your experiment on a chromatography strip that was twice as long as the strip you used, but made of the same material, would your R_f values for the pigments still be the same?

Name ______________________________ Date ______________

7. You work for an art supply company in their marker division. Shown below is a strip of chromatography paper and a table of five ink pigments and their respective R_f values. Mark on the sample strip where each of these colors will be found if mixed together then separated by paper chromatography. Assume the solvent moved 100 units of distance from the starting line.

Pigment Color	R_f Value
Red	.75
Blue	.23
Green	.68
Pink	.94
Purple	.05

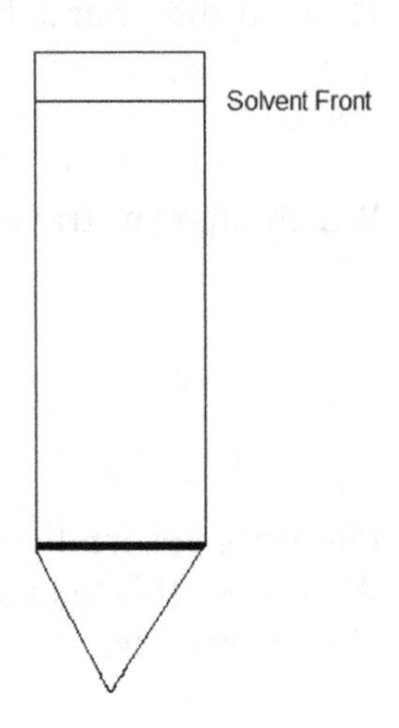

Name ______________________________ Date ______________

Data Table

Distance Traveled by Solvent Front	mm

Pigment Color	Distance Traveled By Pigment	R_f Value
	mm	
	mm	
	mm	
	mm	

Attach your group's chromatography strip here or use this outline to draw a representation of the data of your strip in this location.

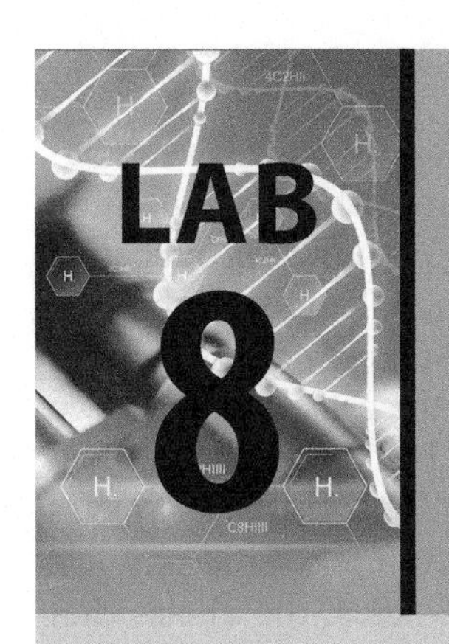

Mendelian Genetics and Probability

BACKGROUND

Gregor Mendel was a German-speaking Augustinian monk who studied the inheritance of traits in pea plants. Gregor Mendel applied mathematical rules of probability to his biological observations which was revolutionary at his time. Most assumed that living organisms would not follow these predictable mathematical laws, but Mendel's carefully designed experiments showed the power of these statistical rules in determining the expected progeny (offspring) frequencies from genetic crosses. Gregor Mendel published his work in 1866 in an obscure German botany journal and his work did not make a big impression on the scientific community of his time. While becoming an abbot in the church, Gregor Mendel died in almost complete scientific anonymity. However, his work was rediscovered 40–60 years after his death, and at that time the researchers realized the significance of Mendel's work, and from that time forward Mendel was considered to be the father of a new field of study, the field of Genetics. Although Mendel worked on garden pea plants, the principles apply equally to humans. Mendel's principles allow us to understand patterns of inheritance of diseases that can be passed down from parent to offspring.

So what were these landmark discoveries that led into a new era of studying inheritance? They can be summed up in two laws: The law of segregation and the law of independent assortment. Mendel always started his genetic crosses with true-breeding lines of pea plants, which turned out to be essential for explaining his results. A line of pea plants would be considered to be true-breeding when they produce offspring with exactly the same traits as themselves when they are self-fertilized. Gregor Mendel would always take his genetic crosses to three generations. The starting true-breeding generation he would call the P generation or Parental generation. The offspring from the two true-breeding lines from the P generation were known as the first filial or F_1 generation. But Mendel would not stop there, he would then take the F_1 individuals and allow them to self-fertilize, forming the second filial F_2 generation.

© Geschaft, 2013. Used under license from Shutterstock, Inc.

The law of segregation states that an individual of a sexually reproducing species has two alleles (copies/gene versions) for each gene. One is received from each parent, because each parent provides one allele in each gamete (sex cell, sperm or egg) they produce. So individuals have two alleles for each gene while gametes have one allele for each gene. This concept is very important when it comes to explaining the results of genetic crosses. We call cells that have only one allele for each gene haploid and the cells that have two alleles for each gene diploid. Humans, like most animals with a backbone, are diploid and sexually reproducing. The only haploid cells in these individuals are the gametes (sperm and egg). Alleles may be dominant or recessive and this determines what physical traits the organisms will show. When an individual produces gametes, the alleles that individual has will segregate (separate) from each other equally. Since an individual has two alleles for each gene, there are two possibilities, either the two alleles will be the same (the individual will be homozygous for that gene) or the two alleles will be different (the individual will be heterozygous for that gene). The combination of alleles which an organism has is known as its genotype. It is the genotype of an organism that ends up determining it physical traits, known as its phenotype. Because of Mendel's law of segregation, we are able to determine what the genotype and phenotype of the offspring of a genetic cross will be as long as we know the parent's genotype. The true-breeding lines that Mendel always started his crosses with were always homozygous for their particular alleles. A genetic cross between two organisms that are differing in a single trait is known as a monohybrid cross. During sexual reproduction, the offspring are the result of the fusing of two gametes together. These gametes are haploid (containing one allele for each gene) so the offspring will be diploid (having two alleles for each gene). These gametes fuse at random (regardless of which alleles they are carrying) so to determine the genetic makeup of the offspring we first need to determine all of the possible gametes which each parent can produce and all of the possible ways in which those gametes can combine.

One way to do this visually is through the use of a Punnett square. What a Punnett square does is tell us the probability of the genetic makeup of the offspring of a genetic cross. The Punnett square does not tell us how many offspring will be made or the order

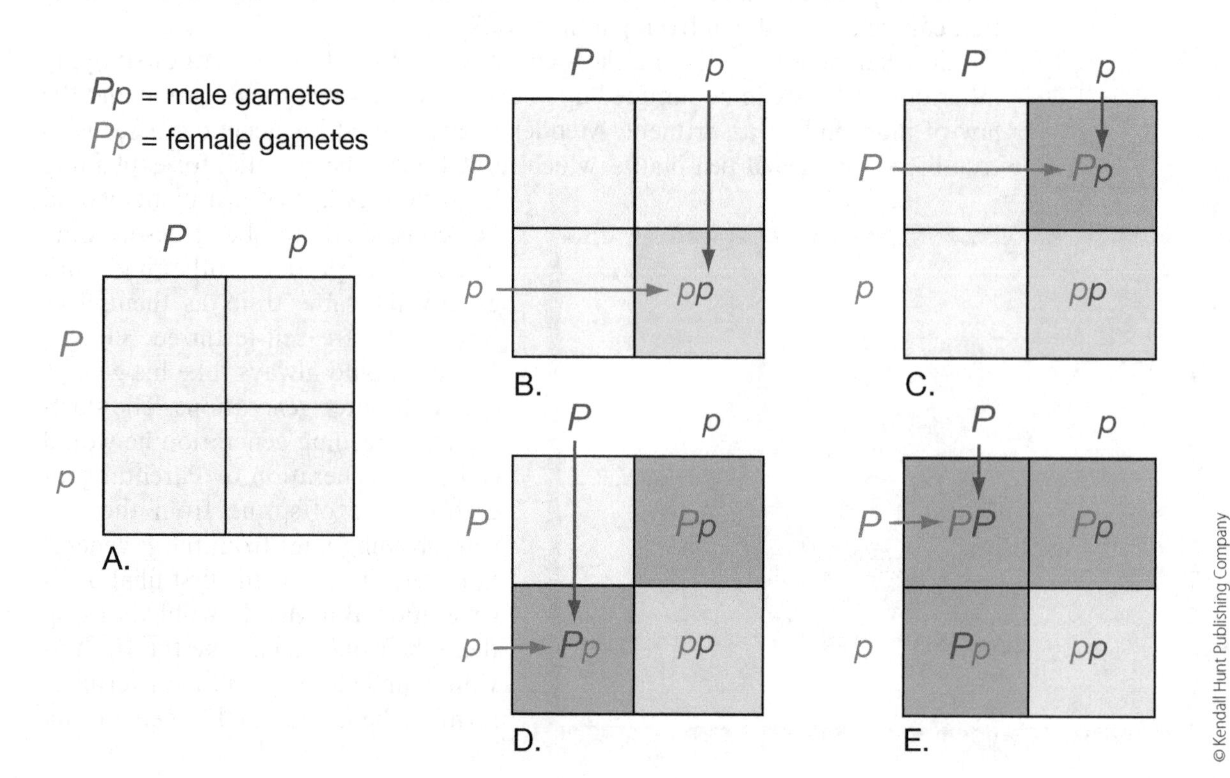

© Kendall Hunt Publishing Company

in which the offspring will develop, but it does tell us, for each offspring that is made, what probability that offspring has for having a particular genotype/phenotype. Each box in a Punnett square represents an equal probability for the offspring's genotype when all the gametes are produced in equal frequencies, so you can simply determine how many boxes in the square have the genotype of interest and that number out of all of the boxes in the square tells you the probability of getting that result in the offspring.

To set up a Punnett square, you need to determine all of the potential gametes of the first parent and place each different gamete type above the columns of the square. Remember the gametes will have only one allele for each gene. Once the different gametes from the first parent are placed above each column, the different gametes from the second parent need to be placed in front of each row. We will be left with a box that will show all of the possible combinations of gametes and by filling in the individual boxes by taking the gametes above each column and in front of each row and combining them within this Punnett square, we will be able to determine the genotype and phenotype of the offspring. By convention and for reasons of clarity, we always write the dominant alleles before the recessive alleles when filling in the boxes of a Punnett square, so when we have a capital and lowercase allele of the same letter (for the same gene) we then always show the capital letter first, whether it came from the top gamete or the side gamete.

Mendel's law of independent assortment has to do with the inheritance of alleles of more than one gene. When we are performing a cross between individuals who differ in two different genes or traits, it is called a dihybrid cross. What Mendel found with his observations was that when an organism is producing gametes, the allele that the gamete receives for one gene does not usually influence the allele that that same gamete will receive for a different gene. The alleles for the two genes assort into the gametes independently of each other. As an example, let's say that we have an organism that is heterozygous for two genes, the A/a gene and the B/b gene. The genotype of this organism would be AaBb. Now when this organism produces gametes, the law of segregation tells us that each gamete will receive one allele for the A/a gene and one allele for the B/b gene. If the gamete receives the A allele; that does not guarantee that it will also receive the B allele. Just as likely, the gamete that receives A could also receive the b allele. The allele that the gamete gets for the A/a gene does not determine what allele is received for the B/b gene. The four possible allele combinations for the gametes would be AB Ab aB and ab, and Mendel's laws of segregation and independent assortment say that all four of them would be found in roughly equal frequencies.

To form a Punnett square for a dihybrid cross, we follow exactly the same rules as for setting up a Punnett square for a monohybrid cross. The keys to remember are that each of the different possible gametes produced by the first parent would be found above a column on the square and each of the different possible gametes produced by the second parent would be found in front of each row of the square. There are a few conventions that we use that may at first seem to contradict what we did with the monohybrid, but are similar and done for reasons of clarity. When working with two different genes, the alleles for the first gene are always listed before the alleles of the second gene in the gametes, regardless of whether they are capital or not. Consistency is important in listing the gametes, the alleles for the first gene (A/a alleles) always need to be listed before the alleles for the B/b gene above each column and in front of each row. When filling in the boxes which represent the offspring, the alleles for the first gene (in this case the A/a gene) will always be listed first in the boxes within the larger square and the alleles for the second gene will always be listed after the first (in this case the B/b gene). This rule supersedes the rule of capital letters always going first. Once we have the A/a alleles together and B/b alleles together (one allele for each gene from the top of the column and one allele for each gene from in front of the row) we then organize each gene by placing the capital letters first then the lower case, but this is done only within each gene. (Examples of proper allele ordering include AaBb, aaBB, Aabb, and AaBB)

	AB	Ab	aB	ab
AB	AABB	AABb	AaBB	AaBb
Ab	AABb	AAbb	AaBb	Aabb
aB	AaBB	AaBb	aaBB	aaBb
ab	AaBb	Aabb	aaBb	aabb

The resulting Punnett square shows us all of the possible offspring genotypes when both parents are AaBb. If an offspring has at least one capital letter for the A/a gene and the B/b gene, it will show the dominant phenotype for both traits. If the offspring has at least one A allele but two b alleles, it will show the dominant phenotype for the first trait but the recessive phenotype for the second. If the offspring has two a alleles, but at least one B allele, it will show the recessive phenotype for the first trait and the dominant phenotype for the second trait. The only way that the offspring will look recessive for both traits is if it received a and b alleles from both parents. What this Punnett square is telling us is that when both parents are AaBb, there is a $\frac{9}{16}$ chance that the offspring will show the dominant phenotype for both traits, a $\frac{3}{16}$ chance that the offspring will show the dominant phenotype for the first trait, but the recessive phenotype for the second trait, a $\frac{3}{16}$ chance that the offspring will show the recessive phenotype for the first trait and the dominant phenotype for the second trait, and a $\frac{1}{16}$ chance of showing the recessive phenotype for both traits. Each box in a Punnett square represents an equal probability when all the gametes are produced in equal frequencies.

It may not be immediately obvious, but the Punnett square is helping you to perform math to estimate the probability of a particular outcome in the offspring. The Punnett square is often the most convenient way of doing the calculation, but there are exceptions. For example, the Punnett square for a trihybrid cross would have 64 boxes and a Punnett square for a tetrahybrid cross would have 256 boxes. When the Punnett square is so large, using the mathematical rules is more convenient. The two rules needed to estimate genetic probabilities are known as the Rule of Multiplication and the Rule of Addition. The Rule of Multiplication states that the probability of a compound event is the product of the separate probabilities of the independent events. You use the multiplication rule when multiple events must occur; this is what is meant by a compound event. One way to determine if you must use the rule of multiplication to calculate a probability is to listen for the word "AND" in your logic statement. If event 1 must occur AND event 2 must occur, then you must multiply the probability of the two separate events to determine the probability of both events actually occurring. The Rule of Addition states that the probability of an event that could occur in multiple different ways is the sum of the separate probabilities of each different way. You use the addition rule when only one of multiple different scenarios must occur. The way to determine if you must use the rule of addition to calculate a probability is to listen for the word "OR" in your logic statement. If you could get your result if scenario 1 occurred OR if scenario 2 occurred, then you need to add those probabilities together to find the probability of either one of them actually occurring.

There are several different ways to represent the probability of an event occurring. One way that probabilities can be represented is as fractions (i.e., $\frac{1}{2}$ or $\frac{3}{4}$). Interpreting these fractions lets us know the number of times to anticipate a particular event in a given number of chances. When flipping a standard two sided coin, $\frac{1}{2}$ of the time the coin is expected to land on heads and $\frac{1}{2}$ of the time the coin is expected to land on tails. This probability can also be represented by percentages. 50% of the time the coin is expected to land on heads and 50% of the time the coin is expected to land on tails, when flipped. There is yet a third way to represent probabilities, and that is through ratios. The ratio of the coin landing on heads or tails is a 1:1 ratio. We would anticipate that the coin would land on heads as many times as the coin lands on tails when flipped multiple times. These probabilities which are calculated are not guarantees. Each time a random event occurs, the outcome is not known beforehand. All that can be known is the probabilities of the particular outcomes. Chance plays a big role in determining the outcome. If a coin is flipped five times in a row and each time it lands on heads, the sixth flip of the coin still has a 50% chance of landing on heads and a 50% chance of landing on tails. Previous results do not affect the probabilities of future outcomes. The more times an event is repeated, the more likely the final results are to agree with the calculated probabilities. That is why using large sample sizes returns results more similar to the expected ratios.

PROTOCOLS

Activity 1: Calculating the Probability of Compound Events

Materials needed by each lab group

2 plastic quarters (one marked with a spot on each side)
1 plastic cup

1. The purpose of this experiment is to see the results of flipping a pair of coins 100 times. (Note: You must actually flip the coins 100 times, don't just flip the coins ten times and multiply your results by ten, you will not be seeing the actual results of probability by doing that.)
2. The result of the coins must be recorded together for each flip by making one mark for the pair in the data table of Question 2, the marked coin determining which column the mark goes in and the unmarked coin determining which row. (Note: The result "• heads, tails" is different than "• tails, heads". It will be helpful to have one or two members flipping the coins, another member marking the tally sheet, and another member keeping track of how many times the coins have been flipped total so that you know when you are approaching the 100 mark.)
3. After flipping the pair 100 times, tally the marks in each box and compare your results with that of a neighboring group.
4. Answer the questions on the student worksheet.

Activity 2: Observing the Effects of Dominant and Recessive Alleles

Materials needed by each lab group

2 Green Plastic Disks
2 Yellow Plastic Disks

1. The purpose of this activity is to see the effect that dominant alleles have on determining an organism's phenotype.
2. Stack the two green disks together to represent a pea plant that is true breeding (homozygous) for green pod color, a dominant trait.
3. Hold that stack of two disks (each disk representing an allele) up to the light and record what color you see shining through.
4. Stack the two yellow disks together to represent a pea plant that is true breeding for yellow pod color, a recessive trait.
5. Hold that stack of two disks up to the light and record what color you see shining through.

6. Now take one yellow disk and one green disk and stack them together. This represents the offspring from a cross between the two true-breeding lines. Each parent contributes one allele to the offspring.
7. Hold this new stack of two disks up to the light and record which of the two colors you see shining through and which of the two colors you do not see.
8. Answer the questions on the student worksheet.

Activity 3: Understanding Independent Assortment

Materials needed by each lab group

2 4-sided dice
1 plastic cup

1. After Gregor Mendel understood the patterns of inheritance for a single gene, he then wanted to know if alleles from two different genes would always be inherited together in a gamete, or if they would be assorted into gametes independent of each other. We will be demonstrating the expected results by rolling pairs of four-sided die to see what the independent assortment of alleles into gametes would produce.
2. For this demonstration, we will be looking at two of the genes that Mendel observed, the gene for plant height (T allele = dominant, tall; and t allele = recessive, short) and the gene for seed shape (R allele = dominant, round; and r allele = recessive, wrinkled).
3. The number on each die represents a specific type of gamete made by a parent pea plant that is (TtRr). Independent assortment tells us that gametes will be produced in every possible combination in roughly equal frequencies, the same way that we should roll the numbers one-four in roughly equal frequencies using a four-sided die.
4. Each number on the die refers to a gamete with a specific allele make up. #1 means the gamete has the TR alleles. #2 means the gamete has Tr alleles. #3 means the gamete has tR alleles and #4 means the gamete has tr alleles. (Note: Remember that gametes are haploid, having only one allele for each gene, but the offspring are the result of two gametes fusing, therefore the offspring will have two alleles for each gene when two haploid gametes fuse.)
5. Roll the two dice together and note the two numbers on the top of the dice together. This die result tells us which two gametes combined to form our hypothetical offspring. (Note: At this point you do not need to record your die rolls, but you want to determine the genotype of the offspring that would form from those two gametes fusing. For example, let's say that the first die came up as a "2" and the second die was a "1." The two gametes are Tr and TR according the instructions above. When these combine, the offspring will have the following genotype, TTRr. If you have filled out your Punnett square correctly for question 16, you can easily find the genotype of the offspring because the first die can determine which column you are looking at and the second die can determine which row. This way you will easily see the genotype of the offspring. Because the table is symmetrical you will see that the result of 2,4 will give you the same genotype as 4,2.)
6. Once you determine the genotype of the offspring, determine the phenotype of that individual and put a tally mark in the table in question 18. (Note: Keep in mind that you are tallying the phenotypes, not the genotypes. There will only be four different phenotypes.)
7. Repeat steps 5 and 6, 79 more times. (Note: For a total of 80 diploid offspring.)
8. Write your group's data on the board and compare your group's results with that of the rest of the class.
9. Cleanup for this lab involves returning the coins, disks, and dice to the cup and returning the cup to the side of the room.

Name ______________________________ Date ______________

Mendelian Genetics and Probability—Student Worksheet

QUESTIONS

1. Using the rule of multiplication, what is the probability of flipping three coins and having all three of them land on heads?

2. Use the following chart to tally the results of your coin flips.

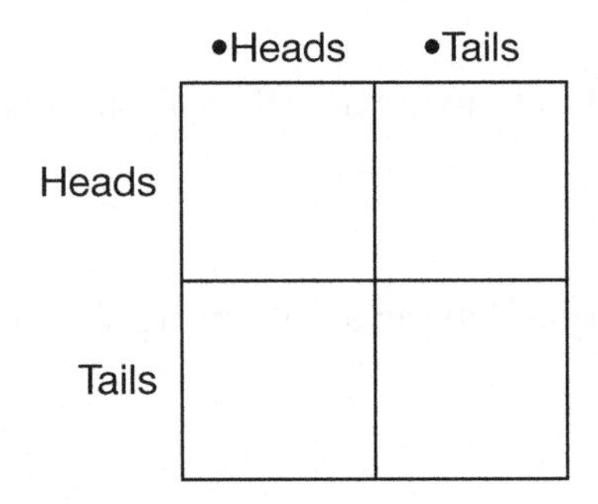

Final Results
•H,H = ______
•H,T = ______
•T,H = ______
•T,T = ______

3. What results were you expecting?

4. How did your results compare with a group near you? Were they exactly the same? Were they similar at all?

5. How would you expect your results would change if you flipped the coins 1,000 times instead of 100 times?

Name ______________________________ Date ______________

6. Complete the Punnett square for the cross between Aa × Aa.

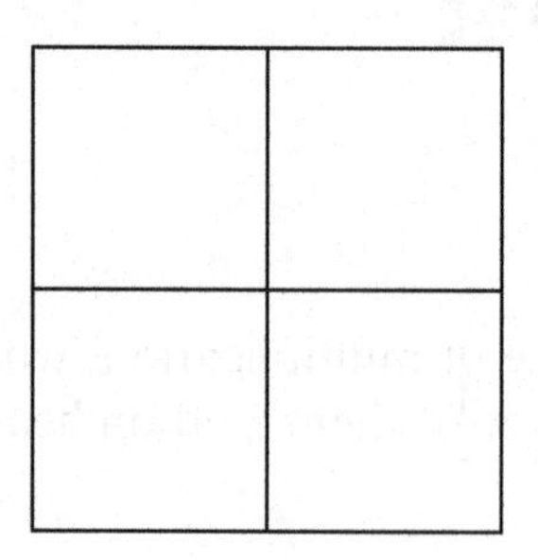

A. How many different types of offspring could be produced by this cross?

B. What is the probability of having an offspring that is heterozygous?

C. What is the probability of having an offspring showing the recessive trait?

7. Complete the Punnett square for the cross between Aa × aa.

A. How many different types of offspring could be produced by this cross?

B. What is the probability of having an offspring that is heterozygous?

C. What is the probability of having an offspring showing the recessive trait?

8. What term do we use to describe the genotype of an organism that is true-breeding?

Name ______________________________ Date ______________

9. When both alleles (plastic disks) were green, what color was visible through the disks?

10. When both alleles (plastic disks) were yellow, what color was visible through the disks?

11. When one allele was green and the other was yellow, what color shone through the two disks? Which of the two colors was not visible through the disks?

12. What does it take to have the yellow color shown in an offspring?

13. Why did Mendel's F_1 generation always only show the dominant trait, but the recessive trait always showed up in $1/4$ of the offspring in the F_2 generation?

14. What does this say about the presence of the recessive allele in the F_1 individuals?

15. Gregor Mendel crossed a true-breeding line of pea plants that were tall and had round seeds (TTRR) with a true-breeding line of pea plants that were short and had wrinkled seeds (ttrr). Both tall plants and round seeds are dominant traits compared to short plants and wrinkled seeds. What would the Genotype and Phenotype of the offspring of this cross be?

Genotype: ______________ Phenotype: ______________

16. Imagine taking one of the offspring from question 15 and allowing it to self-fertilize. Fill out the following Punnett square assuming that the alleles for the two genes assort independently into gametes.

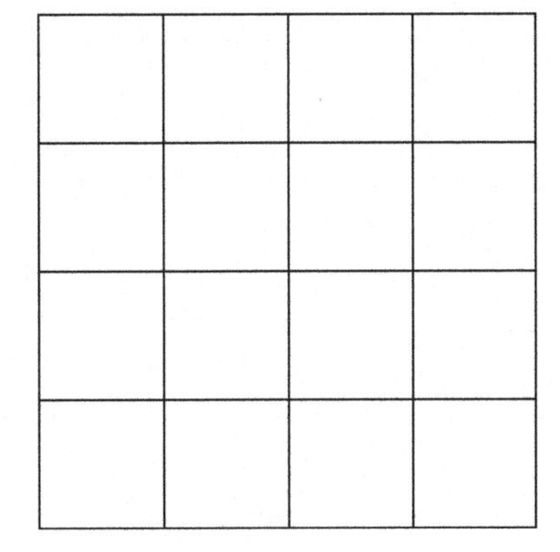

Name ______________________ Date ______________

17. What phenotypes would you expect to be present in the F_2 generation (the results of the above Punnett square)?

18. Tally the results of the die rolls for the four possible phenotypes.

Phenotype	Number in 80 Pairs of Dice Results	Class Average Data

19. How do the results from your group compare with the results from the other groups in the class?

20. Comparing your data and the class data, which is closer to a 9:3:3:1 ratio? Why do you think that is?

21. Gregor Mendel saw a 9:3:3:1 ratio among his F_2 generation in his dihybrid crosses. Why did this convince him of his law of independent assortment?

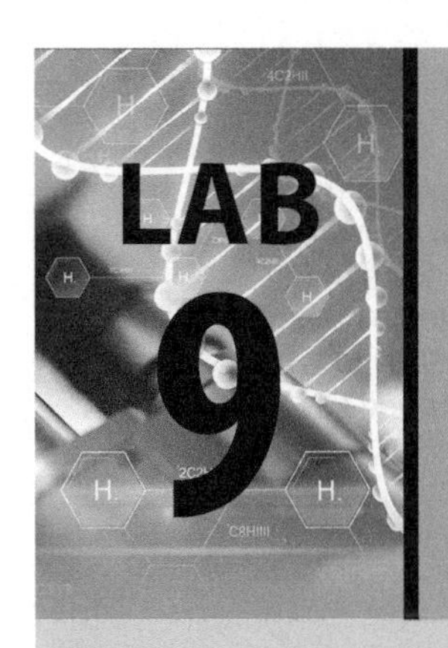

LAB 9 DNA Extraction from Strawberries

BACKGROUND

How do cells function like a factory? Why are we all different? What happens when someone has cancer? Why do people get viral infections? The answers to these questions and many more require understanding of the structure and function of DNA.

DNA (**D**eoxyribo**N**ucleic **A**cid) is the primary information-carrying molecule of the cell. This molecule contains the "blueprints" that the cell needs to make all of its protein machinery along with the production schedule and required quantities of those proteins. The differences that we see between individuals of a species and also the difference that we see between different species is the result of their different protein makeups, which in turn is caused by the differences of the DNA in those organisms. This molecule (DNA), which is capable of storing incredibly complex information, is actually quite a simple macromolecule polymer, being composed of four different types of monomer building blocks called nucleotides. These nucleotides are composed of three parts: one phosphate group, a sugar molecule (deoxyribose for DNA and ribose for RNA), and one of four nitrogenous bases (A, T, C, or G for DNA and A, U, C, or G for RNA). Connecting millions of nucleotides together in long chains results in very long thin thread-like molecules. These long DNA molecules are then wrapped around a protein scaffold to make a structure called a chromosome. Eukaryotic cells may have a few chromosomes or up to over 100 chromosomes, while prokaryotic cells only have a single chromosome. Eukaryotic cells store their chromosomes in an organelle called a nucleus while prokaryotic cells have their chromosome dissolved directly into their cytoplasm. Being an acid, DNA releases H^+ ions when dissolved in water and thus is also a polar molecule. DNA in a cell is always going to be found in a double-stranded state, meaning that there will be two long strands of nucleotides connected to each other through hundreds of thousands, even millions, of hydrogen bonds. While DNA molecules are soluble in water, they will come out of solution if cold alcohol is added to

© maminez, 2013. Used under license from Shutterstock, Inc.

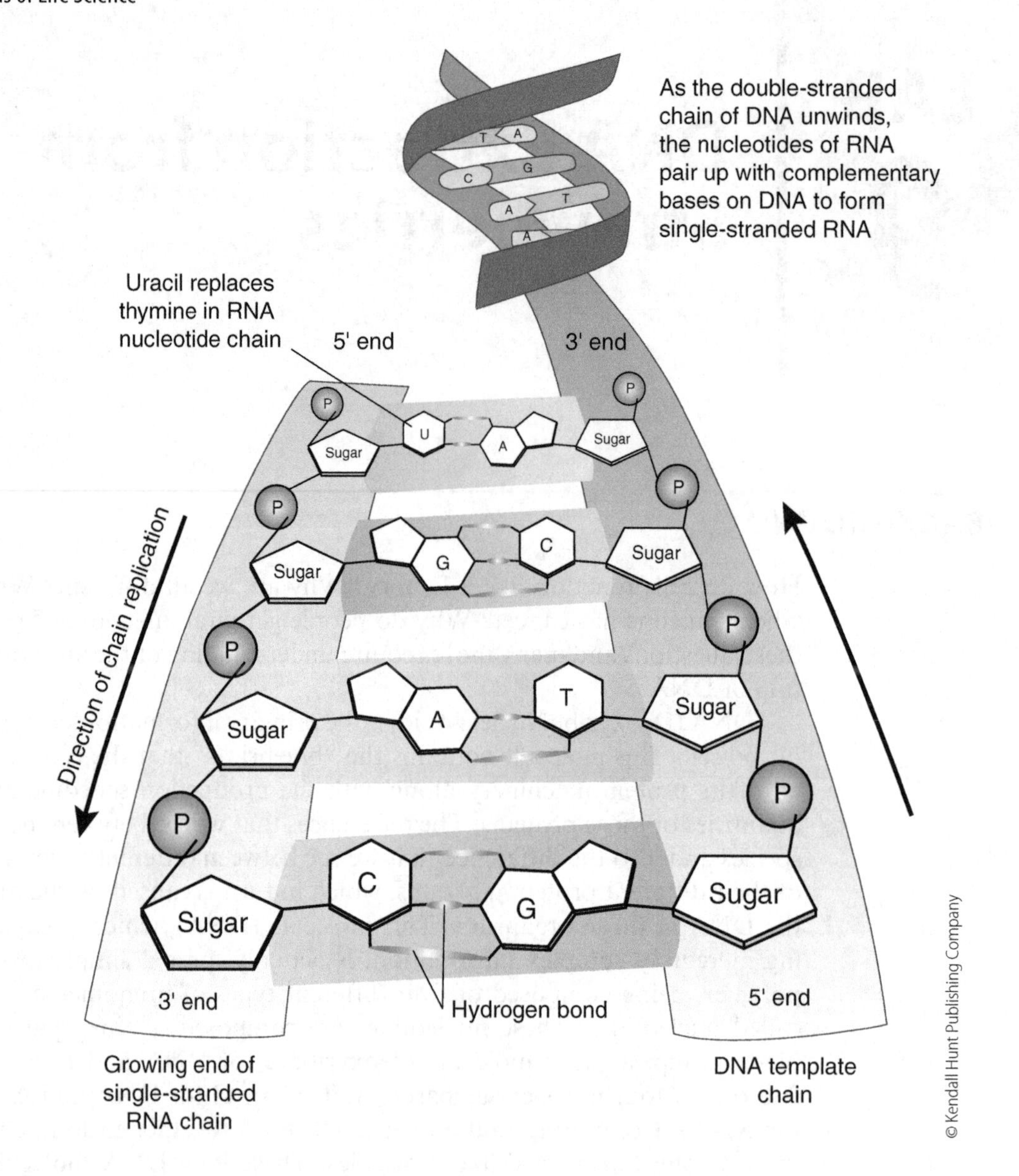

the solution in which the DNA is dissolved. This process of precipitation (coming out of solution) can be used for purifying DNA from a solution containing many other types of molecules. This technique will be used in today's lab to isolate the DNA from strawberries.

Before we talk about precipitation, we need to get the DNA dissolved into an aqueous solution which is accessible to the alcohol. Any solution in which water is the solvent is called an aqueous solution. When we look at strawberry cells, like all plant cells, there are many barriers separating the DNA within the nucleus from a solution outside of the cell. The first barrier which must be overcome is the cellulose cell wall. Cellulose is a strong, nearly-indigestible polysaccharide which all plant cells use as their cell wall. We will need to physically break down this cell wall by smashing the pieces of strawberry using our fingers. In order to keep the contents of the strawberry cells contained along with keeping our hands and clothes clean, the smashing of the strawberries will occur within a plastic ziplock bag. The next barrier after the cellulose cell wall is the plasma membrane of the cell. All cells have a plasma membrane composed of a phospholipid bilayer. This double layer of lipid molecules serves as a barrier which separates the interior contents of the cell from the external aqueous solutions. These membranes allow certain molecules to pass through unhindered while other molecules will be completely blocked by this membrane. DNA is one of those molecules which would be blocked by

© zimmytws, 2013. Used under license from Shutterstock, Inc.

a phospholipid bilayer, so we must break down that barrier if we hope to get DNA dissolved into our aqueous solution. If you have ever seen an oily cooking pan filled with water after a meal, you have likely seen that oil spread as a thin film over the surface of the water. The plasma membrane is similar to this thin film of lipids over an aqueous solution. You have likely also seen the effect of a single drop of soap on this lipid layer, the drop of soap almost instantly breaks this single layer into many smaller globs of oil. We can use this property of soap to help with releasing the DNA trapped within the cell. Eukaryotic cells not only have a plasma membrane, but also contain a nuclear envelope which is made up of not one, but two phospholipid bilayers. The chromosomes of a eukaryotic cell are contained behind this nuclear envelope, within the nucleus. Soap will be essential to breaking down these barriers in order to release the DNA.

Another thing that is holding the DNA within the cell is a protein scaffold. Much of the protein's structure is held together by electrical attraction between the amino acids of the polypeptide chain. By adding many positively and negatively charged ions to a solution containing proteins, the proteins will often lose much of their structure and release whatever other molecules they may have been attached to. Chromosomes are structures composed of double-stranded DNA wrapped around a protein scaffold. By causing the protein to lose its structure, the DNA of the chromosome will be released to dissolve into the surrounding aqueous solution. This is why we will be adding a lysis buffer to the strawberry tissue after our initial round of smashing. The lysis buffer is made of three ingredients: water, salt, and dish-washing soap. The water serves as the solvent allowing all of the soluble components to be dissolved into the mixture. The salt, when dissolved into water, will ionize, forming both sodium (Na^+) and chloride (Cl^-) ions which will help to break down the structure of the protein scaffold of the chromosomes and other intracellular protein structures as well. The soap will have the role we mentioned earlier, of dispersing the phospholipid layers that would otherwise keep the DNA contained within the cellular structures.

The result of mixing the strawberry tissue with this lysis buffer is that most of the cellular contents of the strawberry will now be released and dissolved into the water of the lysis buffer and cytoplasm. This mixture will include proteins, sugars, pigments, DNA, and other assorted molecules. There will also be many components of this mixture that will not dissolve into the solution: the seed-like fruits of the strawberry (known as achenes), along with much of the cellulose from the cell walls. By passing our mixture through a filter (a common coffee filter or a few layers of cheese cloth will work just fine), the components that are dissolved in the solution will pass right through the filter while the larger insoluble components will be left behind. The solution that passes through the filter is where we will find much of the DNA, but at this point, it is still just one of the components of this mixture, as evidenced from the red color and sweet fragrance of the solution. It is at this point that the application of ice-cold ethanol will cause the DNA to precipitate, coming out of solution in such a way that it can be physically removed and observed.

PROTOCOL

Activity: DNA Extraction from Strawberries

Materials needed for each lab group

3–4 slices of strawberry
5 mLs of Lysis Buffer
10 mLs of ice-cold 95% ethanol
1 resealable plastic bag
1 15 mL conical test tube
1 10 mL graduated cylinder
1 funnel
1 coffee filter or a few layers of cheesecloth
1 wooden stirring stick

1. Obtain 1 resealable plastic bag and 3–4 slices of strawberry, placing them into the plastic bag. (Note: When obtaining the slices of strawberry, keep in mind that scalpels are designed for cutting tissue and your hands are made out of tissue, so please be careful!)
2. Every member of the lab group should spend ~ 30 seconds smashing the strawberry slices within the bag, using your fingers. You will improve your chances of success if you keep foam to a minimum during the smashing process. In this process we are physically breaking down many of the cellulose cell walls that are surrounding the strawberry cells.
3. Using the graduated cylinder, measure 5 mL of Lysis buffer (which is made of water, salt, and soap) and pour that volume of lysis buffer into the resealable bag with the smashed strawberry mixture, then rinse the graduated cylinder with water. Again, each member of the lab group should spend an additional 30 seconds each mixing and mashing the strawberry mixture with the lysis buffer. (Note: Consult the background reading to understand what function each component of the lysis buffer is performing.)

4. Place a coffee filter or cheesecloth into a funnel and place the end of the funnel into a 15 mL conical test tube. Then pour the strawberry and lysis buffer mixture into the filter in the funnel. The material that passes through the filter is considered to be dissolved into the solution, while the insoluble components remain behind in the filter. (Note: Once the mixture is poured into the filter, you can fold the edges over and apply gentle pressure to filter the solution more rapidly; be sure to not apply too much pressure as the filter could rupture or the mixture could squeeze out the top.)
5. Collect at least 2 mL and no more than 5 mL of the strawberry solution into the test tube, then discard the filter and residual strawberry mixture, along with the resealable bag. Rinse the funnel and place it next to the sinks to dry. (Note: Some bubbles in the filtered solution are okay, but you want to avoid excessive amounts of foam.)
6. Bring your test tube containing the strawberry solution to the station at the back of the room where the ice-cold 95% ethanol solution is found. Measure 10 mLs of ethanol and add it to the test tube. (Note: The ethanol is denatured and should not be consumed. Hold the test tube at a 45° angle when initially pouring the ethanol into the tube. Two distinct layers will result, with the more dense strawberry mixture at the bottom of the tube.)
7. After adding the alcohol to the strawberry solution, allow the test tube to sit for five minutes undisturbed. (Note: During this five minute waiting period, you may notice bubbles forming and white strands becoming visible at the interface between the two liquids; that is the DNA coming out of solution.
8. Using the wooden stirring stick, reach the interface of the two solutions and gently twirl the stick to draw the DNA to it, then lift the stirring stick from the solution, bringing the DNA with it.
9. Make your observations and answer the questions on the Student Worksheet.
10. Clean up for this lab involves discarding the wooden stirring stick along with the DNA into the garbage can, pouring the strawberry and alcohol mixture down the sinks, and rinsing the test tubes and graduated cylinders with water and returning them to where they were obtained, along with the funnels which were drying by the sink.

Name ______________________________ Date ______________

DNA Extraction from Strawberries— Student Worksheet

QUESTIONS

1. It is important that you understand the steps in the extraction procedure and why each step was necessary. Each step in the procedure aided in isolating the DNA from other cellular materials. Match the procedure with its function:

PROCEDURE	FUNCTION
A. Filter fruit slurry through coffee filter	__ To precipitate DNA from solution
B. Mixing of fruit with salty/soapy solution	__ Separate the insoluble components of the cell from the solution.
C. Initial smashing and grinding of fruit with fingers	__ Break open the cell walls
D. Addition of ethanol to filtered extract	__ Break up proteins and dissolve cell membranes

2. In particular, what purpose do you think the soap served in the lysis buffer? (Hint: What is soap typically good at breaking down?)

3. When you used the wooden stirring stick to pull up the DNA, what did the DNA look like? Please be as descriptive as possible (texture, color, odor, etc. . . . but do not taste or eat). Relate this to what you know about the chemical structure of DNA. Why could you not see a double helix?

4. Considering the strawberry mixture that passed through the coffee filter, the DNA was not visible in the aqueous lysis buffer solution. What does this tell you about the polarity of DNA as a molecule? Do you suspect DNA is polar or nonpolar and why?

5. You have just gone through a DNA extraction process that is typical in many research laboratories. Why might scientists want to be able to remove DNA from an organism? What might they do with it? List two reasons, and be specific (do not just say "they would study it," try to think of what in particular they would want to know from the DNA).

6. DNA is found in pretty much any natural food that we eat. What do you imagine happens to the DNA in our food once we consume it?

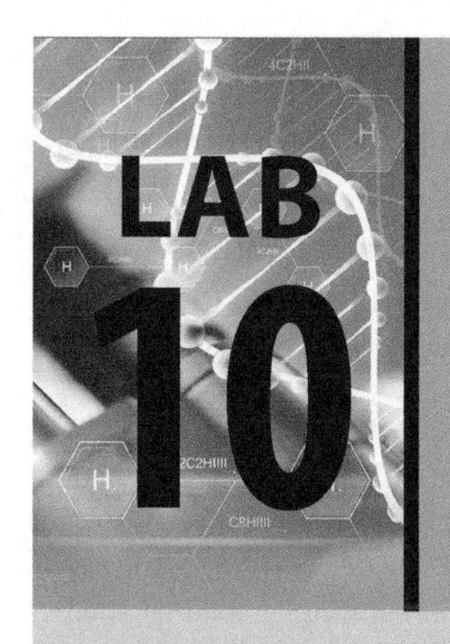

The Central Dogma of Genetics

BACKGROUND

DNA is the genetic material; meaning DNA stores the information that a cell needs to form its proteins. The proteins which are produced, along with environmental factors, eventually determine the phenotype or appearance of an organism. The flow of information from the DNA through the RNA to the Protein is known as the Central Dogma of Genetics. The process of transcribing RNA molecules from a DNA Template is known as Transcription. The DNA sequence of the gene will determine the sequence of the mRNA molecule which is produced from transcription. The mRNA molecule will be complementary to the DNA template and its sequence can be determined by following the complementary base-pairing rules. Remember that DNA has the bases A,T,C, and G; and A pairs with T and C pairs with G. RNA has the bases A,U,C, and G; and A pairs with U and C pairs with G. If you have a DNA/RNA Hybrid, A on the DNA will pair with U on the RNA, all of the other base pairing will be the same. The mRNA molecule is the result of transcription, and is complementary to the DNA template strand. The mRNA molecule is then transferred to the cytoplasm where it is translated to form a polypeptide chain with a specific amino acid sequence (protein). The molecule responsible for translating the mRNA sequence into an amino acid sequence is tRNA. tRNA has two binding sites, the anticodon (which is complimentary to the codon sequences of the mRNA) and the amino acid binding site. The tRNA molecule translates each codon of the mRNA into a specific amino acid, except for the stop codons, which signal the end of translation. Unlike life on Earth, the fictitious organisms we will be studying in this lab do not have start or stop codons marking the beginning and ending of their mRNA molecules.

PROTOCOL

1. In this lab activity, you will be given the template DNA sequence of a chromosome for a fictitious organism. From this DNA sequence you will need to determine the complementary mRNA sequence.
2. You will use the mRNA sequence to determine which tRNA anticodon sequences will be complementary to the mRNA.

3. By using the chart on the front of the worksheet, you will be able to determine which amino acids correspond to which anticodon sequences.
4. Each gene codes for a protein that will be three amino acids long. There is also a chart on the front of the worksheet which tells you what allele each protein codes for. Once you know the traits of your organism, you then must draw a picture of it.

Name ______________________________ Date ______________

The Central Dogma of Genetics

Instructions: In this activity, you will examine the DNA sequence of a ficticious organism: the Snerp. Snerps were discovered on the planet Kro-Mos-Ome in a neighboring solar system. Some of the traits of Snerps are controlled by their six genes on their single chromosome, their other traits are controlled by their environment. Your job is to analyze the DNA of a specific individual and determine what traits the organism has, due to its genetics.

Snerp DNA and Traits

tRNA Triplet	Amino Acid Number
ACC	20
AGC	16
CGA	2
AAC	4
CGC	3
GGG	5
AGG	7
AAA	8
UUU	9
GGU	12
UAU	13
CCC	1
AUC	6
CUA	10
GGA	11

Amino Acid Sequence	Trait
20–11–13	hairless
20–12–13	hairy
20–4–4	plump
13–4–4	skinny
16–2–5	four legged
16–4–5	two legged
12–7–8	round head
5–7–8	block head
9–8–8	no tail
9–4–8	tail
11–3–2	triangle eyes
11–3–3	circle eyes
6–6–10	male
6–6–1	female

OBSERVATIONS AND ANALYSIS OF SNERP DNA

You are given the template strand of a chromosome from a Snerp with the following sequence. Each gene codes for a protein that has only three amino acids. Your job is to determine the sequence of amino acids for the six proteins of your organism which determines its traits. Write the complementary mRNA, tRNA, the amino acid (A.A.) sequence it codes for, and the related trait in the following chart. Then draw a picture of the organism in the provided space.

Name ______________________ Date ______________

DNA	ACCGGTTAT|AGCCGAGGG|TTTAACAAA|GGACGCCGA|GGGAGGAAA|ATCATCCTA
mRNA	
tRNA Anticodon	
A.A.	
Trait	

Draw your Snerp in the space below. Be creative!

Name ______________________ Date ______________

DNA	ACCGGATAT|AGCAACGGG|TTTAAAAAA|GGACGCCGC|GGTAGGAAA|ATCATCCCC
mRNA	
tRNA	
A.A.	
Trait	

Draw your Snerp in the space below. Be creative!

Name ______________________________ Date ______________

DNA	A C C G G T T A T | A G C A A C G G G | T T T A A C A A A | G G A C G C C G C | G G G A G G A A A | A T C A T C C C C
mRNA	
tRNA	
A.A.	
Trait	

Draw your Snerp in the space below. Be creative!

Name ______________________ Date ______________

DNA	ACCGGATAT|AGCCGAGGG|TTTAAAAAA|GGACGCCGA|GGTAGGAAA|ATCATCCTA
mRNA	
tRNA	
A.A.	
Trait	

Draw your Snerp in the space below. Be creative!

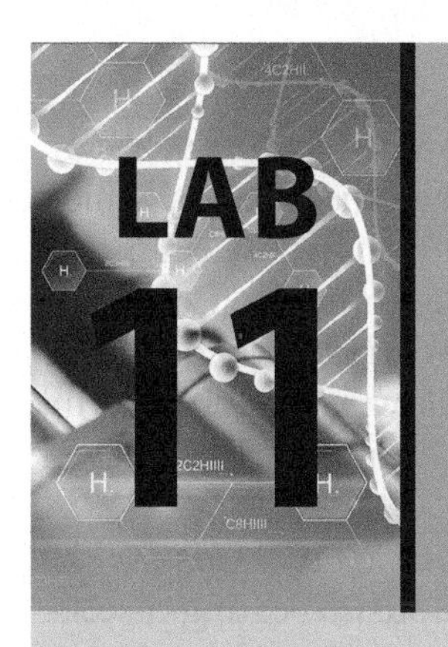

LAB 11 Bird Beak Adaptation

BACKGROUND

Natural selection can be defined as a process of differential reproductive success due to being better suited to one's environment. What that means is that certain organisms will produce more offspring than others because of their specific physical traits and characteristics. This in turn means that their offspring (which will have some genetic similarity with their parents) will make up a larger proportion of the next generation, changing the characteristics and genetic makeup of that population. This is what the phrase "survival of the fittest" refers to. We often think that "fitness" means muscle strength or endurance or stamina, but the biological definition of this term "fitness" refers to reproductive success. This reproductive success is not just measured as the number of offspring that are produced, but the number of offspring that survive to produce their own offspring. While strength and stamina may help an organism compete for food or space or other resources that allow it to be more reproductively successful; strength and stamina on their own are not biological fitness. Those organisms that are best suited to their environment and best able to compete for limited resources are usually those that will be most reproductively successful.

Charles Darwin developed the Theory of Natural Selection based on his research of finches in the Galapagos Islands. He postulated that all species of finches on the islands descended from the same original species; however, due to natural occurrences, the finches were eventually isolated on the individual islands. This type of isolation is known as geographic isolation. The isolated finches on each island slowly diverged from the original species and adapted to their specific environment. Each new species of finch took on the characteristics needed for survival in its own unique niche. For example, Darwin noticed that despite a strong resemblance to the original species, each new species of finch had a highly characteristic beak shape, adapted to the specific environments to which these finches were found.

© ranker, 2013. Used under license from Shutterstock, Inc.

Copyright © 2002 by Ward's Science. Reprinted by permission.

Because of the adaptations they made to their different environments, the finches on each island became reproductively isolated from the finches on the other islands, which in turn led to speciation. Darwin noticed that the birds had only themselves to compete with for food. This competition caused the birds, in this struggle, to develop different habits for survival. Birds that developed specialized traits, by evolving, would survive and reproduce, causing different characteristics to emerge, and allowing the birds to fill different niches. For example, through natural selection, the seed-eating finches evolved a thick, bulky, cone-shaped beak for cracking open seeds.

Among each island, however, there was still competition among many birds for the same resource. For example, more than one type of bird was adapted to eating insects. And, more than one type of bird was adapted to eating fish. This competition for resources is apparent in all environments. One way in which species reduce this direct competition is by resource partitioning. Resource partitioning is known as the dividing up of resources so that species with similar needs use them at different times, in different ways, or in different places. For example, some birds will feed on insects at night while others will feed during the day. In effect, they evolve traits that allow them to share the wealth.

According to Darwin's Theory of Natural Selection, individuals with the best combinations of inherited traits are the most likely to survive and reproduce. This means that over time, populations of animals with specific adaptations to a particular environment are more numerous than populations without these specific adaptations. Animal adaptations can be any body shape, process, or behavior that allows an organism to survive in its environment, and they may change over time to fit the needs of the environment. Birds can be found throughout the entire world, and they display a wide range of adaptations depending on their environment. Beaks are used for eating, defense, feeding young, gathering nesting materials, building nests, preening, scratching, courting, and attacking. The shape and size of each species' bill is specific for the type of food it gathers. For instance, a pelican has many unique adaptations, including a spoon-like beak, that make it better at catching fish than a woodpecker. In turn, pelicans can be found in aquatic environments where fish are abundant. Woodpeckers, on the other hand, would not be able to survive in an aquatic environment because they are not adapted to that particular environment. The different beak shapes allow easier access to particular food supplies. So, if an environment is altered, organisms within the area

© attaphong, 2013. Used under license from Shutterstock, Inc.

will need to change, or adapt, in order to survive. Natural selection is the process by which organisms best suited to the environment survive and reproduce, thereby passing their genes on to the next generation.

Animals which form some sort of social structure, including humans and birds, may benefit from learning by watching others in their group. For example, chimpanzees learn how to dispose of parasites from watching others in their group do so. Many birds learn songs from parents or others nearby. Humans learn language from interaction with other humans, not from some innate knowledge provided by their genes. Such learning can change the shape of natural selection for a particular population. For example, chimpanzee populations that use sticks to harvest ants can rely more heavily on ants as a food source rather than other foods. This means that characters that were advantageous for harvesting other foods, such as being a fast runner, may suddenly become less important in the population. So inventors may be important in more than just human societies! While you are conducting your study on bird beaks, be alert for opportunities to use a seemingly inappropriate 'beak' to harvest the resource in new and interesting ways, and pass the information along to others in your group.

PROTOCOL

Scenario

You and your classmates will represent four different types of birds. As a group, you will visit four different islands. Upon closer inspection, you notice that each island contains a different type of food for birds. It is your mission to explore how each bird beak can be used to obtain food on each of the four islands. Each test will allow you to see how different bird beaks are adapted to their food source.

1. Your teacher has set up four islands. Each island has a unique environment, and in turn, has a unique food source for birds:
   Island #1: Aquatic Vegetation
   Island #2: Worms
   Island #3: Seeds
   Island #4: Nectar
2. Your class will be divided into groups containing at least four people.
3. Begin with any of the four islands. As a group, note the food present. The 'aquatic vegetation' island will be corks floating in the water, the 'worms' island will be some pipe cleaners in potting soil, the 'seeds' island will be a plate of sunflower seeds, and the 'nectar' island will have containers of water. Record this information in Table 1 below.
4. Read the rules for the food item you will be harvesting below, then hypothesize what a bird would have to do to obtain the particular food source. Record your hypothesis in Table 1.
5. Predict which "beak" on this island will be the most effective in obtaining the food. Every station will have four beaks: a pair of pliers, a pair of forceps, a dipnet, and a pipet. Record your prediction in Table 1.
6. Discuss the reasons why you believe this beak will function most effectively. Record your explanation in Table 1.
7. After you have completed the preliminary information on the island, a member of your group should raise his/her hand to indicate that the group is ready for the competition. Instructions for the competition appear below.
8. Repeat steps 3-7 for each island, switching after you have completed the competition at the current island.

General Overview of the "The Best Beak Competition"

- Each type of beak will be tested on each island by each group.
- This will be a competition to see *how much* of each particular food source can be "consumed" within 15 seconds using each beak. Remember that you are trying to survive! You can use any part of your tool to get the resource into your cup, so long as you follow the rules for each island detailed below.
- Your teacher will be the timer, using a stopwatch to time 15 seconds for each trial.
- Three trials will be performed using each beak, and then an average of the three trials will be calculated in order to determine the average amount of each food consumed.
- Group members should play a different role for each of the three trials so we do not bias our results. For each trial, decide upon a role for you and your group members using the following:
  - Birds: Four members of your group will try to obtain food using a tool. Each person may use only one tool per 15 second trial, and a person cannot use the same tool at the same station during any of the three trials.
  - Replenishers: If you have more than four group members, the others should assist in measuring the food consumed and restoring the station for the next trial.
  - Reader: One group member will read the rules at each island and make sure that all group members understand.
  - Recorder: One group member will record the correct values in the chart on the observation sheet.

Hint: For each trial, group members can rotate in and out of these roles as they desire, so each group member has a chance to participate. Also, if there are not enough group members to perform all of these role, some group members can take on more than one role at a time.

- On each island you may begin with any beak of your choice.
- You may visit the islands in any order. Just make sure to mark the appropriate information for each island in the proper spaces on your Tables.
- Your teacher will collectively time each competition on each island. Always have one member raise his or her hand when your group is ready for a new trial to avoid unnecessary delays in running the trials.

9. Carefully read the rules below for each island so you know the object of the competition and how to measure the amount of food "consumed".

Rules for Island #1: Aquatic Vegetation

- Using one hand to operate each beak, you have 15 seconds to "consume" as many pieces of aquatic vegetation from the pond as possible. Your teacher will tell you when to begin and when to stop each trial.
- Place each "consumed" piece of vegetation into the empty cup to be counted after each trial. After the pieces are counted, they should be returned to the water for the next trial.
- Record your counts and calculate the resulting averages to complete Table 2.
- Perform three trials using each beak, replacing the vegetation in the pond after each trial.

Rules for Island #2: Worms

- Using one hand to operate each beak, you have 15 seconds to "consume" **one at a time,** as many worms from the soil as possible. Your teacher will tell you when to begin and when to stop each trial.
- Place each "consumed" worm into the empty cup to be counted after each trial. After counting, the worms should be placed back into the soil for the next trial. While placing the worms back into the soil, make sure that the people who are acting as birds in the next trial are not looking. It needs to be a random feeding.
- Record your counts and calculate the resulting averages to complete Table 2.
- Perform three trials using each beak, replacing the worms in the soil after each trial.

Rules for Island #3: Seeds

- Using one hand to operate each beak, you have 15 seconds to CRUSH ("consume") **one at a time,** as many seeds as possible. Your teacher will tell you when to begin and when to stop each trial.
- Place each "consumed" seed into the empty cup to be counted after each trial. After counting, the crushed seeds should be placed in the trash. Do not place crushed seeds back into the pan.
- Record your counts and calculate the resulting averages to complete Table 2.

Rules for Island #4: Nectar

- Using one hand to operate each beak, you have 15 seconds to "consume" as much nectar from the flower as possible. Your teacher will tell you when to begin and when to stop each trial.
- Make sure that the water is filled to the top line before each trial.
- Dispense all "consumed" nectar into the empty cup. Measure it by carefully pouring it into one of the graduated cylinders. After the liquid is measured, pour it back into the 'flower' container and add additional water to reach the top of the container. It is very important that you empty the cup after each trial.
- Record the volume collected for each trial and calculate the resulting averages to complete Table 2.
- Perform three trials using each beak, replenishing the nectar to the top line after each trial.

11. Once your data is complete, you can determine which beak was best suited for the food on each island. Record your decision in Table 2.
12. Discuss the structural advantage of the best beak on each island and why it was the best, in the appropriate spaces in Table 2.
13. Clean up materials as follows:
    Aquatic vegetation: Remove corks to a paper towel, pour water down the sink, then replace corks in the plates. Use paper towels to soak up any additional water.
    Worms: Replace all of the 'worms' back into the soil. Rinse out the tools and cups or plates which got dirty; dispose of the pipette. Use paper towels to wipe the table and place excess soil in the trash.
    Seeds: Dispose of all broken or cracked seeds in the trash. Stack plates containing remaining seeds and set aside. Use a paper towel to wipe debris from the table.
    Nectar: Pour all 'nectar' down the sink. Use paper towels to dry the table.

Name ______________________________ Date ______________

Bird Beak Adaptation—Student Worksheet

DATA TABLES

Table 1—Predictions

Island	Type of Food	What will a bird have to do to obtain this food?	Which "beak" will work the best in obtaining the food?	Why?
Island #1				
Island #2				
Island #3				
Island #4				

Name ______________________ Date ______________

Table 2—Observations

Island	Beak Type	Trial 1	Trial 2	Trial 3	Average	Which "beak" worked the best?	Describe the structural advantage	Examples of birds having this beak structure
Island #1: Aquatic Vegetation	Pipet							
	Pliers							
	Tweezers							
	Dipnet							
Island #2: Worms	Pipet							
	Pliers							
	Tweezers							
	Dipnet							
Island #3: Seeds	Pipet							
	Pliers							
	Tweezers							
	Dipnet							
Island #4: Nectar	Pipet							
	Pliers							
	Tweezers							
	Dipnet							

Name ______________________________ Date ______________

QUESTIONS

1. Define adaptation.

2. Describe three ways in which you are adapted to your environment.

3. There are many natural and man-made changes to the environment that can affect different species of birds. Choose two of the following changes to an environment:
   - Deforestation
   - Insecticide application
   - Drought
   - Aquatic oil spill
   - Climatic change (e.g., snowfall in an area where it is atypical)

   Describe how each change would impact the everyday lives of birds in the area. Give an example of an adaptation that would allow the bird to be better suited to its new environment.

4. Choose one island that you visited. Construct a bar graph using the averages from each beak that depicts the relative effectiveness of each beak. Make sure to label your graph properly. Explain what the graph is demonstrating in terms of natural selection.

Name ______________________ Date ______________

5. Was there any evidence of direct competition on any of the islands (i.e. were any beaks comparable in effectiveness)? Give two examples of how different species might be able to reduce the direct competition for the same food source.

6. The island model in this activity has limitations. For example, it demonstrates competition for only one endless food supply on each island. Imagine if this food supply was destroyed. What survival strategies could birds employ if their food source was destroyed?

7. Imagine that you are a hummingbird. You have a very specialized beak that allows you to extract the nectar from flowers. What benefit does your specialized beak provide? What would be the consequences if your food source disappeared through a natural catastrophe?

8. Did you have any situations where you learned to use a tool in a way that you had not originally thought to harvest the resource? If so, explain, including what island, what tool, and what new method were involved.

Biochemical Evidence for Evolution

BACKGROUND

All of our evidence for evolution suggests that evolution is a remodeling process, taking structures and molecules that are present and changing them into new shapes and new molecules with new functions through the process of random mutation of the genetic material. This is how new alleles for genes are generated. The more similarly-shaped certain molecules and structures are, the more closely they are related and the more recently they diverged from the same common molecule or structure. This is the theory behind the fields of comparative anatomy and comparative molecular biology. We call this type of similarity homology. As with all things in biology, there are a few exceptions which serve to complicate matters, but for the most part this finding is consistent. The more recently structures shared a common origin, the more similar they will be in structure.

We will demonstrate this similarity in today's lab, which is a simulated agglutination reaction. Agglutination is a clumping reaction which is the result of a protein from a vertebrate's immune system, known as an antibody, binding to a specific molecule which it recognizes, called an antigen. Each antibody has two or more binding sites for its antigen and so the result when both antibodies and antigens are found together is that they will begin clumping; the agglutination reaction which was mentioned earlier. Antibodies have a very specific shape at their antigen-binding site and will only bind to an antigen with a complementary shape. The more complementary the shape, the better the antibody will bind to the antigen. Antibodies are generated when an antigen is introduced into the body fluids of a vertebrate. An organism will only produce antibodies specifically for those antigens that it has been introduced to and that are not naturally produced by that organism itself. For the commercial production of antibodies, the antigen is often introduced into a rabbit and as that organism produces antibodies, those antibodies are collected. We can use these antibodies and their specificity to determine the similarity in shape of a protein from several different organisms. This is what we will be analyzing in a simulated manner in today's lab.

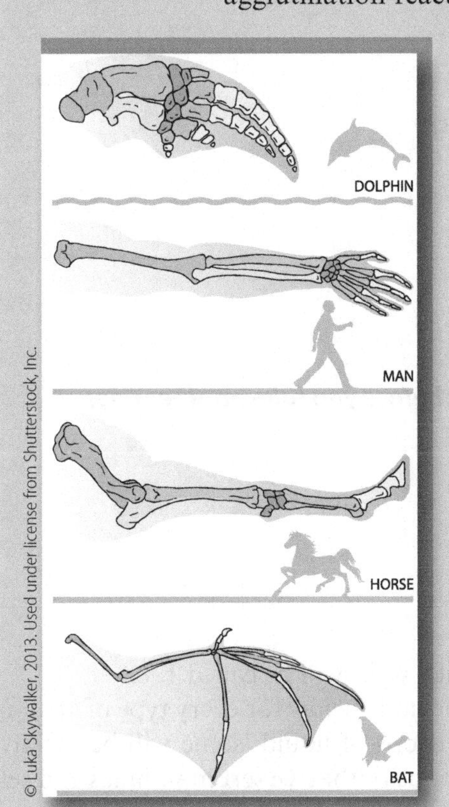

© Luka Skywalker, 2013. Used under license from Shutterstock, Inc.

A general blood protein known as albumin was isolated from humans and introduced into rabbits. These rabbits

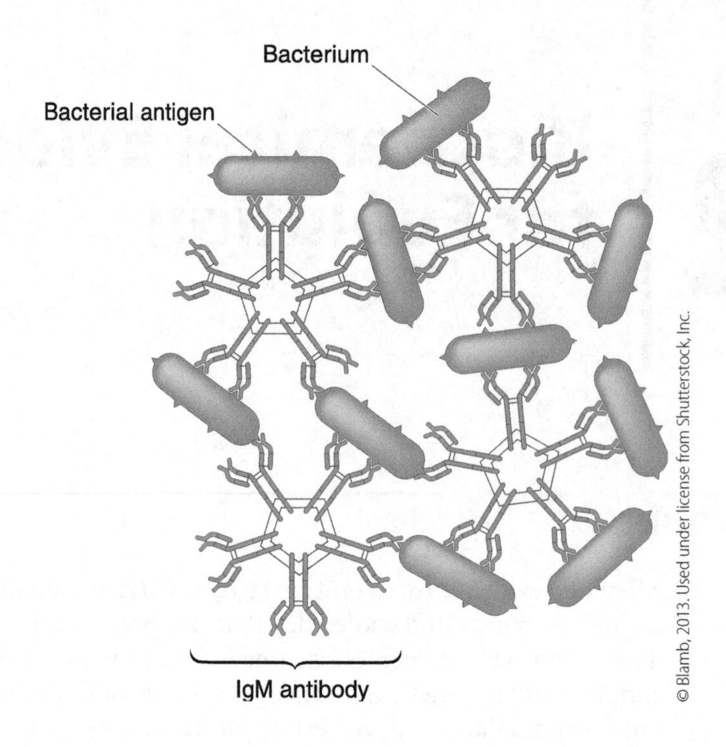

then produced antibodies against human albumin that did not match any antigens produced by the rabbit itself. Those antibodies can then be tested on samples of albumin from several different organisms (chimpanzees, rhesus monkeys, cows, and frogs) including human albumin, and we can compare the agglutination reactions between these various samples. The more similar in shape the molecules of the sample are to human albumin, the more pronounced the agglutination reaction will be.

PROTOCOL

Activity: Determining Molecular Homology

Materials needed by each lab group

1 or more multi-welled sample plate
several stir sticks
Human Antiserum

Shared materials (Use one at a time, then return it before you take another one)

Human Serum
Chimpanzee Serum
Monkey Serum
Cow Serum
Frog Serum

1. Start by taking a sample plate and deciding which wells will be called 1–8.
2. You will be doing the dilution series shown in the table below for every type of serum. Notice that each well will contain a total of 8 drops of liquid; some will be mostly water, while others will have more serum. Select one of the five serum samples and set up the following dilution series in the labeled wells of your plate.

Well Number	Drops of Serum	Drops of Water
1	8	0
2	7	1
3	6	2
4	5	3
5	4	4
6	3	5
7	2	6
8	1	7

(Note: After this dilution series is set up, you should have a total of eight drops of liquid in each well.)

3. After you have filled all eight wells with the appropriate combination of serum and water, add an additional eight drops of human antiserum in each well. Stir each well, starting with well 8 and working towards well 1, rinsing the toothpick after each full cycle. (Note: With the addition of the antiserum, there will be a total of 16 drops of liquid in each well. The reason we stir the wells from 8 towards 1 is that we are moving against our dilution gradient so we do not need to rinse the toothpick between each well because we are going from less concentrated solutions toward more concentrated solutions. If you move from 1 to 8 without rinsing the toothpick between each well, there is a risk of contaminating your results.)
4. Allow the agglutination reaction to continue for two minutes, then record the intensity of the agglutination in each well on the data table using a scale of 0–3:

Agglutination category	Description
0	No clumps visible on surface or in solution
1	Very few clumps visible on surface
2	Clumps quite frequent across the surface
3	Surface choked with clumps throughout

5. After recording your results, rinse off the tray and clean out all of the wells used in this round, dry off the tray, then select a different serum and repeat the process. Following the same procedure from steps 2–4 for all five different sample sera.
6. Clean up for this lab involves rinsing the trays and stirring sticks one final time, taking care to scrub off residue from your work, and returning all of the shared materials to where they were obtained.

Name ______________________ Date ______________

Biochemical Evidence for Evolution—Student Worksheet

QUESTIONS

1. Fill out the following data table based on the results of your experiment using the 0–3 scale.

Organism	Well 1	Well 2	Well 3	Well 4	Well 5	Well 6	Well 7	Well 8
Human								
Cow								
Chimpanzee								
Frog								
Monkey								

2. Based on the data you collected, which of the organisms are most related to humans, and which the least? Create an evolutionary tree showing how the different organisms are related. (Note: At the bottom of the tree are the ancestors and branches which separate from that base as new species are formed. Since we are using human antiserum in this experiment, humans should be at the top of the tree and those organisms with the greatest agglutination reactions even when diluted would be closer to humans and those with the weakest agglutination reactions farther away. See question 5 for an example of an evolutionary tree.)

3. Why do we include human serum as one of the samples that we test in this experiment?

4. Why couldn't we use rabbit serum as one of our tested samples and expect to get accurate results? Be specific. Would you expect too much or too little agglutination with rabbit serum and our human antiserum?

Name ______________________________ Date ______________

5. Below is an example of an evolutionary tree involving a deer, pig, and beaver. If a new organism were discovered, the "snickerloo," which had a pig's head, antlers, and a large deer-like body, where would you place it on the evolutionary tree? Draw on the tree below. The place where your new line originates is your hypothesis about how closely related the snickerloo is to the other groups. The line can originate off any of the lines in the tree. For example, if we thought the snickerloo was most closely related to a beaver, we might have the snickerloo branching off from the line leading to the beaver.

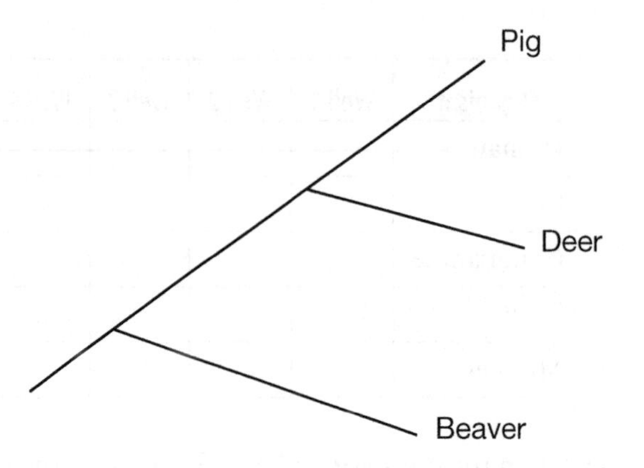

6. If you could not rely on anatomical structures (for example, if this snickerloo also has a flat, beaver-like tail, making it difficult to place on the tree provided). Describe how you would set up a test similar to the one we did today in lab in order to find the snickerloo's relationship to these other animals. Be sure to mention which animal you would make the antibodies against and what the results of this experiment would tell you about their relationship.

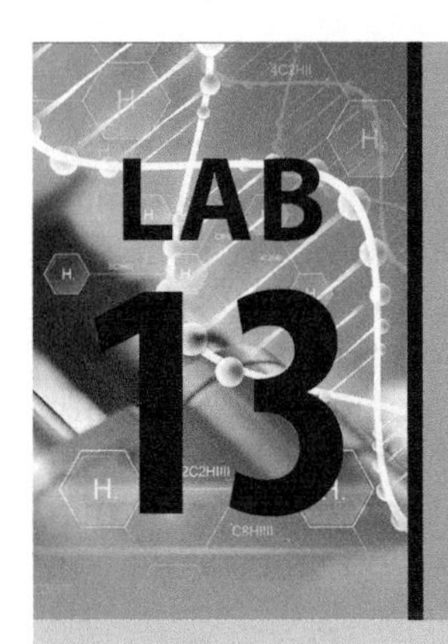

LAB 13 Human Blood Typing

BACKGROUND

Blood is a connective tissue that is composed of a combination of several types of cells suspended in a liquid extracellular matrix called plasma. Both white and red blood cells are present in our blood, but the red cells (erythrocytes) outnumber the white cells (leukocytes) 700 to 1, so that is why our blood is red in appearance. These erythrocytes are covered with molecular signature molecules of various sorts. We call these signature molecules antigens. All cells have these signature molecules that identify what type of cell they are. Our immune system is constantly screening cells in our body to see if there are any signature molecules that mark a cell as being foreign (non-self). Our immune system produces specialized proteins for this task called antibodies. These antibodies are specific, meaning they recognize a particular antigen and they bind to it. Our body makes antibodies against those antigens that we do not produce on our own. In this way, the immune system makes antibodies that specifically mark or bind to foreign cells.

Back to the antigens on our erythrocytes; there are a few signature molecules that are of note and give humans their blood type. These antigens come in the form of the A and B type sugars. There is a gene known as the I gene that determines an individual's blood type. This particular gene has three different alleles: I^A, I^B and i. The I^A allele codes for the protein attachment for the A type sugar, the I^B allele codes for the protein attachment for the B type sugar and the i allele codes for no sugar attachment site at all. Based on Mendel's law of segregation, we know that each individual will have two alleles for each gene and the I gene is no exception. The different possible combinations of alleles are I^AI^A, I^Ai, I^BI^B, I^Bi, I^AI^B, and ii. Individuals with I^AI^A and I^Ai genotypes would have the A type sugar on their erythrocytes and therefore their phenotype is that they would be blood type A. Individuals with I^BI^B and I^Bi genotypes would have the B type sugar on their erythrocytes and therefore their phenotype is that they would be blood type B.

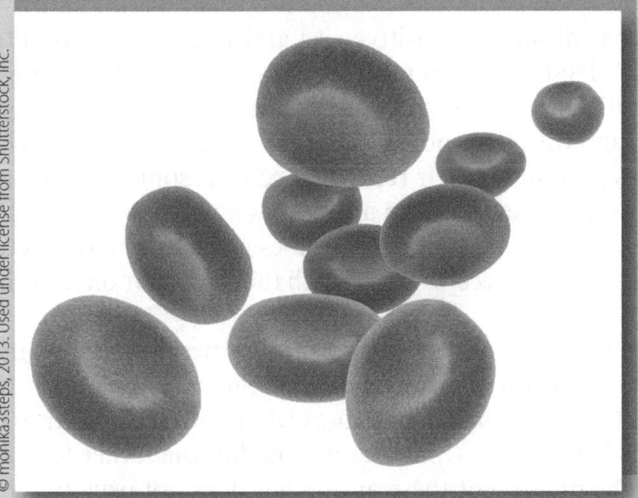

© monika3steps, 2013. Used under license from Shutterstock, Inc.

Individuals with the I^AI^B genotype would have both A and B type sugars on their erythrocytes so their phenotype would be blood type AB. This is an example of codominance, because the heterozygote is fully showing the effect of both of these alleles. The last genotype of interest is ii, these individuals would have neither A nor B type sugar on their erythrocytes, so their phenotype is that they would be blood type O. While the alleles I^A and I^B are codominant with each other, both of them are completely dominant over the recessive i allele. Even though blood type O is caused by two recessive alleles, it is actually the most common blood type in the United States. This just goes to show that just because an allele is recessive, does not mean that is rare. The second most common blood type in the U.S. is A, followed by B and lastly AB.

When we look at how blood type affects the immune system, remember that the immune system will make antibodies against antigens that the individual does not produce on their own. Someone who is blood type A will produce A type sugars on their own, but not B type sugars. This means that someone who is blood type A will produce antibodies against the B type sugar. In that same way, someone who is blood type B will produce antibodies against the A type sugar. Blood type O individuals do not produce A or B type sugars on their blood so they will produce both Anti-A and Anti-B antibodies, whereas individuals with the AB phenotype will not produce Anti-A or Anti-B antibodies. These antibodies become significant when we consider treatments which require individuals to receive a blood transfusion. In these cases, an individual is receiving blood from a source other than themselves. When we look at the types of blood that an individual can receive we need to make sure that they are not receiving blood that contains antigens that their immune system will react to. What happens when blood of an incompatible type is introduced into a patient is that the patient's own antibodies will start binding to those blood cells, causing them to clump together in a process known as agglutination. Agglutination, if it is occurring on a large scale inside the body, could cause severe complications, including death. Someone who is blood type A produces Anti-B antibodies, so could not receive blood of type B or AB, due to the presence of those B type sugars. The person with type A blood could, however, receive type A or type O blood in a transfusion without having an immune reaction to it. A person with type B blood could receive type B or type O blood, but not A or AB due to the presence of the A type sugar. Someone who is type AB could receive any type of blood without an immune reaction, which is why they are known as the universal acceptors, while someone who is type O could only receive type O blood because they produce both Anti-A and Anti-B antibodies, but could donate blood to any other blood type. Individuals who are type O are known as universal donors because anyone could receive their blood without complications.

You may be wondering, "What about the positive (+) and negative (−) that are often associated with blood types?" That is referring to a separate protein antigen known as the Rh factor. Like the A and B type sugars, this is a signature molecule that some people have on their erythrocytes and others do not. If someone produces the Rh factor, they are +, because that protein is present on their red blood cells. If someone does not produce the Rh factor, they are − and they produce antibodies against the Rh factor. Someone who is Rh positive can receive blood that does or does not have the Rh factor, but someone who is Rh negative should not receive blood with the Rh factor on it or they will have an immune reaction.

Current medical policy in the U.S. and other developed countries is that a woman who is Rh negative will receive a Rhogam shot during her pregnancy. This shot will prevent the woman's immune system from producing an excessive amount of Anti-Rh antibodies just in case her fetus is Rh positive (this is done whether the father tests Rh positive or not, just to be sure). The buildup of these antibodies does not usually affect the first pregnancy of an Rh positive fetus in the Rh negative mother, but if the woman were to become pregnant a second time, if that second fetus was also Rh positive, the

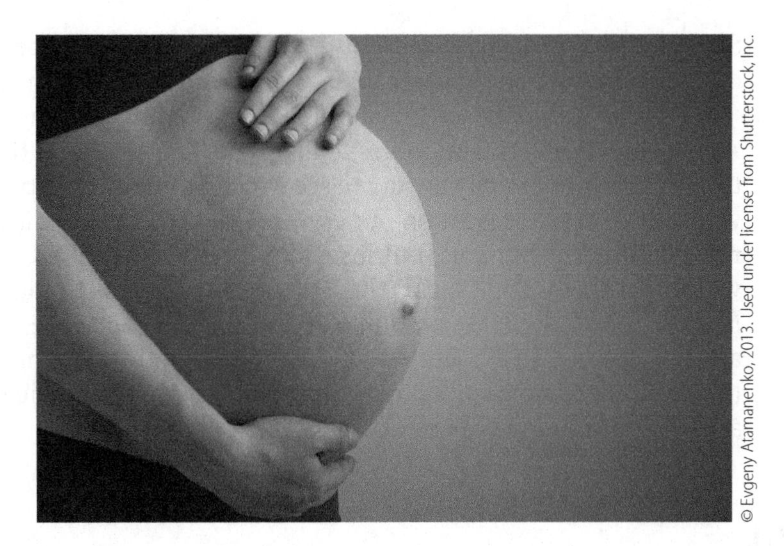
© Evgeny Atamanenko, 2013. Used under license from Shutterstock, Inc.

woman's immune system would respond much more quickly and much more strongly to that second Rh positive pregnancy and the complications of such an immune response could result in damage to the developing fetus and possibly the loss of the pregnancy and developing child. This condition is known as erythroblastosis fetalis, in that the red blood cells (erythrocytes) of the fetus end up getting destroyed by the immune system of the mother. If the mother had not had the Rhogam shot or in some other way had become sensitized to the Rh factor before her pregnancy, one treatment that has been used in the past is to perform a blood transfusion for the fetus in-utero, to eliminate the Rh factors as the fetus completes its embryonic development.

When we look at the antigens across the entire US population, we can see that some blood types are much more common than others:

A-B Blood type	Positive	Negative
O	38%	7%
A	34%	6%
B	9%	2%
AB	3%	1%

(Source: American Red Cross, 2016)

There is some variation that is related to country of origin. For example, 25% of people from Asia are B+, a much higher percentage than we see in the US population at large. Notably, most large ethnic groups that have been tested have every possible blood type.

Researchers have long been interested in correlating blood type with other characteristics related to health and personality. We would see such associations if the ABO or Rh gene is physically close, or linked, to another gene, or if the blood type genes affect more than one trait (called pleiotropy). A few of these studies have observed associations of traits with blood type. A study in 2015 found that people with an A allele (A or AB blood) did not receive as strong a benefit from a prostate cancer therapy compared to people with B or O blood. A separate study of nearly 90,000 people in 2012 suggested that people with type O blood have the lowest risk of coronary heart disease, followed by those with A (5% higher than O), B (11% higher than O), and AB (23% higher than O). Although significant, it's important to maintain perspective in light of these results.

Since the risk of heart disease for O was only 0.37%, a person with AB blood has a 0.52% chance of getting heart disease. Your risk is low with either blood type.

The way we will be testing our blood type in today's lab is by mixing drops of Anti-A, Anti-B, and Anti-Rh antibodies to drops of our blood and then checking for agglutination. We are not directly testing to see which antibodies we produce; we are testing to see what antigens are present on our erythrocytes. The antibodies that we will be using were not produced by humans, but instead produced by rabbits. These rabbits were injected with either the A Type sugar, the B Type sugar or the Rh factor, and this stimulated the rabbits to produce large amounts of antibodies against these foreign substances. These antibodies were then collected to be used in blood-typing procedures. Seeing which antibodies cause agglutination of our blood tells us which antigens are found on our blood and what our blood type is.

PROTOCOL

Activity: Human Blood Typing

Materials needed for each participant

As many gloves as you have hands
1 plastic 3-welled tray
1 lancet
2 alcohol wipes
3 toothpicks
Shared Materials
Anti-A, Anti-B, and Anti-Rh antibody solutions
Biohazard "Sharps" container for lancets
Biohazard bag for any garbage that does/may have blood on it (including trays, toothpicks, gloves, and wipes)
Typical garbage bag for any material that is not at risk of being contaminated with blood.
White piece of paper

1. Of all of the labs that we do this semester, this lab is the most dangerous. Not that losing three drops of blood will in any way cause any lasting harm, but in a class of this size it is possible that an individual could have a disease which would be communicable through blood contact. We have never had any problems with this lab in the past, but it has been because the students have shown the proper respect for minimizing chances of blood contamination throughout the room and by disposing of the garbage in the proper receptacles. For this reason and to retain the ability to continue to perform this lab in future semesters, it is very important to follow the safety procedures outlined in this protocol.
2. Participation in this activity is voluntary. You do not need to participate or type your blood if you have apprehension to doing so for any reason, but each student must completely fill out the Student worksheet and so would need to enter the information of someone else in the class who did get their blood typed to complete certain charts and answer certain questions.
3. Your instructor may direct you to obtain some warm water and soak the finger that you plan to puncture for a few minutes before you come up. Soaking the finger softens the tough, dry skin layers to make the puncture more efficient and reduce the chance of needing multiple punctures. The participants will be invited forward six

at a time for the typing procedure. As you approach the lab bench, obtain a pair of gloves to use during the procedure. Put one of the gloves onto the hand that you will not be puncturing.

4. Opening one of the alcohol wipes, wipe the finger that you will be puncturing. Then discard the wipe and wrapper into the regular garbage, since there is no blood at this point. (Note: I often recommend the side of the finger because there are fewer pain receptors there compared to the tip or flat part of the finger. I also recommend the ring finger on the non-writing hand to minimize interference with other activities throughout the day.)
5. Using the lancet (spring loaded or traditional, depending on availability) puncture the skin, release any pressure at the site of the puncture, then starting at the base of that finger, massage blood toward the puncture site. Discard the lancet into the sharps container, the lancet's wrapper or cap can be discarded into the regular garbage or the biohazard waste if there is any blood on it.
6. As a drop of blood begins to form at the site of the puncture, touch the drop to one of the wells on the plate. Massage blood to the puncture site again and repeat for the other two wells on the plate. (Note: We are not looking for "gravity-drops" of blood; it does not need to fall from the finger on its own. You should not be attempting to fill the wells with blood, a single spot of blood in each well will be sufficient for typing.)
7. After each well has a drop of blood, using the other alcohol wipe, clean the blood from your finger and put the other glove onto the punctured hand. Discard the alcohol wipe into the biohazard waste along with its wrapper if the wrapper has blood on it. (Note: An adhesive bandage is not required at this point; simply put the glove onto the hand, the puncture site will likely clot over the course of the next few minutes of the procedure. If a bandage is needed at the end of the procedure, it will be provided.)
8. At this point, there is nothing different about the drops of blood in each of the different wells, but what we will be adding to the wells will be different. Using the provided dropper, add one drop of the Anti-A antibody solution to the well labeled "A," one drop of the Anti-B antibody solution to the well labeled "B," and one drop of the Anti-Rh antibody solution to the well labeled "Rh_o." (Note: The antibody solution drops should be "gravity-drops" in that we do not want to touch the blood with the droppers in the antibody solution, we want the drops to fall into the wells, to reduce the risk of contaminating the shared components.)
9. With the blood and the antibody solutions together in the well, using a different toothpick for each well, gently mix the liquids together. Discard the toothpick into the biohazard waste because it now has blood on it. (Note: The toothpicks do not need to be put into the "Sharps" container, plastic is not usually considered sharps.)
10. With each well mixed, put the plastic plate on the white paper and assess for agglutination (clumping). (Note: The Anti-A and Anti-B antibody solutions tend to work rapidly and show clear signs of clumping, the Anti-Rh antibody solution is usually slower to react and the reaction begins with the mixture starting to show a granulated or sandy appearance.) You may need to allow the agglutination reaction to proceed for several minutes. Before assessing the result, again gently stir each solution, as the cells can sometimes sink to the bottom, giving the appearance of agglutination. If they can be stirred back into the solution, this is not true agglutination.
11. Once you have determined which wells have agglutination and which do not, you will be able to determine your blood type. Mark your blood type on the class data table on the board.

12. Place the plastic tray, and your gloves into the biohazard waste bag. (Note: If your finger is still actively bleeding, wipe the finger with an alcohol wipe again and then apply an adhesive bandage, be sure to throw the alcohol wipe into the biohazard waste.)
13. Return to your seat and record the data on agglutination on question 1 of the student worksheet.
14. The entire bench will be wiped down before the next group of six participants will be invited forward to type their blood.

Name ______________________________ Date ______________

Human Blood Typing—Student Worksheet

QUESTIONS

1. Fill in the data table using the information from your blood sample or from one of your classmates. For each solution write "Agglutination" if there was clumping in that well or "No Agglutination" if there was not. Enter the blood type based on these results and mark whether this was your blood or not.

Anti-A Solution	Anti-B Solution	Anti-Rh Solution	Blood Type	Check this box if this is your actual sample

2. Based on the above information, which antigens (of the ones we tested for) are found on the surface of the erythrocytes for this blood sample?

3. If the typed individual (you or classmate) needs a blood transfusion, which types of blood could they receive without having an immune response?

4. Based on the above information, which antibodies would you or your classmate's immune system be able to produce?

5. Complete the following table using the class data from the board.

Blood Type	# of Participants with Blood Type	Total # of Participants in the Class	% of Participants with Blood Type

6. Which blood type is the most common in the U.S. and does our class sample match that of the country's population as a whole? Why or why not?

Name ______________________________ Date ______________

7. What is erythroblastosis fetalis and what causes it?

8. You are a medical technologist and you are given the job of determining the blood type of "Patient X". Fill out your report based on the following results.

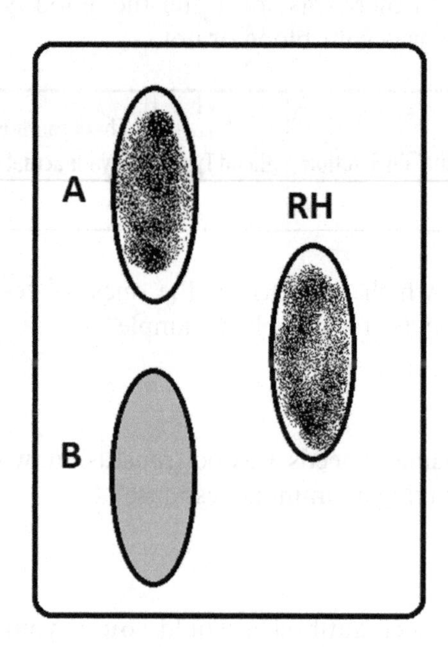

Medical Technologist's Report

Patient Name: ______________

ABO Type: ______________

Rh Type: ______________

Med. Tech. Name: ______________

9. You are a lawyer representing a man with blood type B. A woman of blood type O has a child with blood type A and is proposing that the man fathered the child. Using what you have learned about blood typing and genetics, briefly state your argument to the jury in defense of your client.

10. You are a lawyer representing a woman with blood type A who has a child of blood type O. The alleged father is of blood type B and claims that there is no way that he could be the father because of his blood type. Using what you have learned about blood typing and genetics, point out to the jury the possible flaw in his argument. (Note: Read carefully as this is a slightly different situation than question 9.)

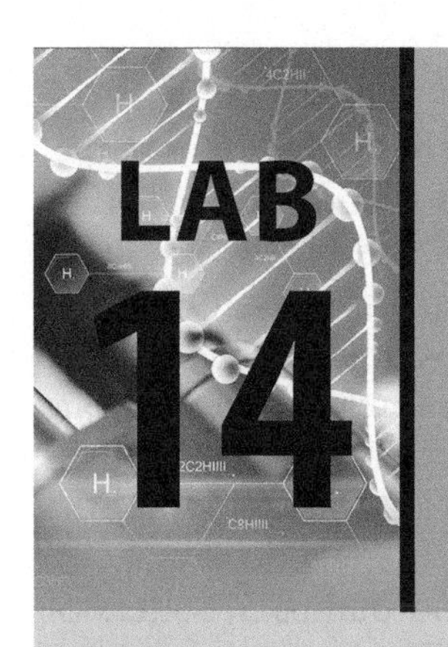

Physiology of the Circulatory System

BACKGROUND

The survival of any organism depends on its ability to establish an internal environment that will keep individual cells alive and healthy. The maintenance of this internal environment in a steady state is called homeostasis. In complex organisms such as humans, homeostasis can only be maintained with a transport system that meets a wide range of needs. The blood, heart, and circulatory vessels carry out the necessary transport function. Contraction of the ventricles of the heart increases blood pressure and thereby forces the blood into the arteries. As the ventricles relax, blood pressure drops. As a result, blood pressure cycles between a high and a low. The highest pressure reached in the cycle is called the systolic pressure and the lowest pressure reached is the diastolic pressure. Blood pressure is expressed as the height in millimeters that it will raise a column of mercury (mm Hg). The systolic pressure is written first and the diastolic pressure second (e.g. 120/80 mm Hg). Baroreceptors located in the carotid arteries and aortic arch constantly monitor blood pressure and send nerve impulses to the brain. The brain sends nerve impulses to the heart, arterioles, and other organs to increase or decrease the blood pressure as needed.

It is standard medical procedure to take blood pressure readings in the brachial artery of the arm, at the level of the heart. In some cases, pressure in the right and left arm can be different in the same person, so often the left arm is used to monitor a person's blood pressure over time to standardize the measurement. Blood pressure is routinely measured with a sphygmomanometer.

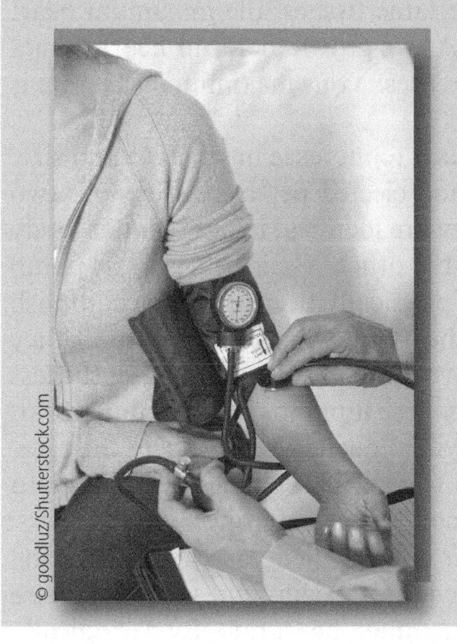
© goodluz/Shutterstock.com

The sphygmomanometer consists of an inflatable cuff, a pump, a gauge graduated in millimeters of mercury, and an exhaust valve with a screw control. The cuff is wrapped around the upper left arm just above the elbow and then inflated. The examiner listens for sounds from the brachial artery by placing the bell of a stethoscope on the inside of the elbow below the biceps. When the pressure in the cuff exceeds that in the artery, the artery collapses and blood flow stops. The pressure in the cuff is allowed to fall gradually by opening the exhaust valve. As the pressure in the cuff drops, it reaches a point at which the pressure of the blood forces the artery open slightly, allowing a turbulent flow of blood to pass. Smoothly flowing blood and arteries that have blood stopped are silent, but turbulent blood makes a sound you can hear through the stethoscope. The turbulence sets up vibrations in the artery that are heard as sounds in the stethoscope (called Korotkoff sounds). When the sound first becomes audible, it is a sharp thumping. The cuff pressure at which the sound is first heard is read as the systolic blood pressure. As pressure in the cuff decreases, the sharp thumping sound becomes louder and then muffles. The cuff pressure at which the sound disappears is read as the diastolic pressure. An organism's blood pressure has to be sufficient to circulate blood throughout the body, particularly against gravity. It takes less pressure to circulate blood on a lateral plane than on a vertical plane.

At rest, a human heart beats about 72 times each minute and pumps 5–6 liters of blood. As a person begins to exercise, tissues need more oxygen. The heart responds by increasing its beats per minute and thus increasing the volume of blood circulated.

© Leremy/Shutterstock.com

Eventually the heart reaches a point at which it is physically impossible for it to fill with blood and contract any faster. This is the maximum heart rate. Because all normal hearts have the same structure and are made of the same tissues, the maximum heart rate is much the same for everyone at the same age. As we grow older, our heart tissues become less elastic, and the maximum heart rate decreases. Your maximum heart rate is approximately 220 minus your age.

Because the heart is largely muscle, activity causes it to increase in strength and size. This increases its stroke volume, the amount of blood ejected per beat. Consider two people running around a track at the same pace. They are identical in age, gender, body mass, and so on, but one is more physically fit than the other. Their hearts must pump the same volume of blood per minute, but the more physically fit heart can accomplish this with fewer beats. As the two run faster, their heart rates will increase until they reach their maximum rates. Although both hearts have the same maximum rate, the less physically fit heart will reach its maximum sooner, because it pumps less blood per beat. Thus, over a distance, the more physically fit heart will allow its owner to win the race. The heart of the physically fit person will also have to beat fewer times per minute to circulate the required amount of blood when the person is not exercising.

Target heart rates are used as a way to pace your efforts when you exercise. Pacing yourself is especially important for sedentary individuals beginning a new exercise program. Your target heart rate is 50–75% of your maximum heart rate. By periodically monitoring your pulse as you exercise and by attaining your target heart rate, you can effectively and safely receive the benefits of being physically active. When beginning an exercise program, aim at the lowest part of your target zone (50% of your maximum heart rate). Gradually build up to the higher part of your target zone (75% of your maximum heart rate). After six months or more of regular exercise, you might be able to exercise comfortably up to 85% of your maximum heart rate if you wish—but you do not need to exercise that hard to stay in good condition.

The circulatory system brings the oxygen and food and removes the carbon dioxide and other waste related to cellular metabolism. The metabolic rate can be quite different between animals that use metabolic energy to maintain their body temperature (called endotherms) and animals that rely on the environment to determine their body temperature (called ectotherms). The metabolism of ectotherms is much slower than for endotherms because ectotherms do not maintain a constant body temperature independent of their environment. From approximately 5°C to 35°C, the rate of metabolism in these animals increases as environmental temperatures increase. Recall that an increase in temperature means that molecules, including proteins, are moving more rapidly than at lower temperatures. When enzymes, which do much of the metabolic work within cells, move faster, reactions occur more quickly, leading to the increase in metabolism. Since metabolism increases, the heart rate increases as well, so that transport of oxygen and other substances moved through the circulatory system will be sufficient to meet the needs of the cells. Endotherms maintain their body temperature at a constant level across a broad range of external temperatures.

PROTOCOLS

Activity 1—Measuring Blood Pressure

Materials needed for each group

Sphygmomanometer
Stethoscope
Alcohol wipes
Timer

Introduction

For this activity, you will work in group of 3–4 and take turns measuring each other's blood pressure using a sphygmomanometer and stethoscope. One of you will serve as the test subject, one as the examiner, one as the data recorder, and one as the timer. The roles of the data recorder and timer can be combined if you have three in your group. Switch roles and repeat the activity so everyone has a chance to measure blood pressure. To give you a point of reference for what you are measuring, the American Heart Association recommends treatment for people with blood pressure over 140/90 (systolic/diastolic).

Note: *These lab results are determined for experimental purposes only. They are not a substitute for regular, professional health care and diagnosis.*

1. Timing is important, so read the instructions before you begin the activity. The test subject should be seated with sleeves (if any) rolled up past where the cuff will fit.
   - Clean the earpieces of the stethoscope with an alcohol swab before and after use.
   - Never leave an inflated cuff on anyone's arm for more than a few seconds.

- Inspect the sphygmomanometer. Be certain that the exhaust valve is open and that the cuff is completely deflated.
- Wrap the cuff snugly, but not tightly, around the upper left arm 2–3 cm above the bend of the elbow.
- Place the bell of the stethoscope directly below the cuff in the bend of the elbow.
- Close the exhaust valve of the bulb (pump) and rapidly inflate the cuff by squeezing the bulb until the pressure gauge goes past 180 mm Hg, or 20 mm Hg above the systolic blood pressure that your subject informs you is typical for him or her.
- Open the exhaust valve just enough to allow the pressure to drop slowly, by about 2–5 mm Hg/sec.
- As the pressure falls, listen with the stethoscope for the first appearance of a clear thumping or tapping sound. The pressure at which you first hear this sound is the systolic pressure. Direct the data recorder to record the systolic pressure in Table 1.
- Continue to listen as the pressure falls. The sound will become muffled and then louder. When the sound disappears, note the pressure. Record this measurement in Table 1 as the diastolic pressure. Open the exhaust valve to completely deflate the cuff.

2. Allow the subject to relax for 30–60 seconds before proceeding.
3. Repeat step 1 two more times for the same subject, to complete trials 2 and 3. Determine the subject's average systolic and diastolic pressures using the values from the three trials. Record all results in Table 1 below.
4. You need to record results for only one individual in your group to turn in.

Table 1—Blood Pressure While Seated. 'Total' is the sum of values for all three trials. Calculate the average by dividing the total by the number of values (3)

	Systolic	Diastolic
Trial 1		
Trial 2		
Trial 3		
Total		
Average		

Activity 2—Testing Physical Fitness

Materials needed for each group

Sphygmomanometer
Stethoscope
Alcohol wipes
Timer

Introduction

Physical fitness involves many components and can be defined in many ways. A champion gymnast, for example, might perform poorly in a marathon. The following tests are chosen to determine the ability of your cardiovascular system to adapt to change. This is one measure of general physical fitness. As you proceed, be alert to signs of dizziness or faintness in the test subject and be ready to steady or catch the subject if you are needed. **Notify your teacher of any medical condition that might make it inadvisable for you to participate in any of these tests.**

You will work in groups of 3–4. One student will serve as the test subject, two as the examiners (one for pulse and the other for blood pressure), and one as the data recorder and timer. If you have three members, one examiner will also serve as data recorder and timer. Familiarize yourself with the procedures before you begin. We will be looking at both blood pressure and pulse rate for this activity. The pulse can be taken at the wrist as shown in the figure, or the pulse can be found by gently pressing on the carotid artery just below the jaw line on the side of the neck.

© caimacanul/Shutterstock.com

Part 1: Systolic Blood Pressure and Pulse: Standing and Reclining

1. The subject should stand at ease for two minutes. During this time, the subject should avoid moving his or her legs.
2. After two minutes have passed, the pulse examiner should count the subject's pulse rate for 30 seconds on the right side. Multiply the rate by 2 to get beats per minute. Record the pulse in beats/minute in Table 2 below the instructions for this activity. At approximately the same time, the blood pressure examiner should measure blood pressure on the left side of the subject. Record the systolic blood pressure in Table 2.
3. The subject should recline for five minutes. You should try to have the subject's heart, head, and legs at approximately the same level. The subject can lie on the lab bench to achieve this position.
4. After five minutes, the two examiners should simultaneously take the subject's systolic pressure and pulse rate. Record the values for reclining blood pressure and pulse rate in Table 2.
5. The subject should remain reclining for two minutes after Step 1 and then stand up with arms down at his or her side. The examiners should *immediately* take the systolic pressure and pulse rate (on the other arm). Record the data in Table 2. **Caution:** *It is possible to become dizzy after standing in this manner. If the test subject becomes unsteady, becomes pale, or complains of feeling faint, seat them at once. Instruct the test subject to lower his or her head between the knees and keep it down until the sensation passes.*
6. Determine the change in systolic pressure by subtracting the reclining systolic pressure from the systolic pressure immediately upon standing. Do the same for the pulse rate. Record the information in Table 2.

Table 2—Systolic Pressure and Pulse Rate Standing, Reclining, and Immediately after Standing

Test	Systolic Blood Pressure	Pulse Rate (Beats per Minute)
Standing		
Reclining		
Immediately upon Standing		
Increase upon Standing (Immediately upon Standing—Reclining)		

7. Look at the value in Table 2 for the systolic blood pressure 'increase upon standing.' Use the table below to assign points based on your result. Circle the appropriate row, and record the points in the first row of Table 4.

Change (mm Hg)	Points
Rise of 8 or more	3
Rise of 2–7	2
No rise	1
Fall of 2–5	0
Fall of 6 or more	−1

8. Look at the value in Table 2 for pulse rate while the subject was standing. Use the table below to assign points based on your result. Circle the appropriate row, and record the points in the second row of Table 4.

Beats/minute	Points
60–70	3
71–80	3
81–90	2
91–100	1
101–110	1
111–120	0
121–130	0
131–140	−1

9. Look at the value in Table 2 for pulse rate while the subject was reclining. Use the table below to assign points based on your result. Circle the appropriate row, and record the points in the third row of Table 4.

Beats/min	Points
50–60	3
61–70	3
71–80	2
81–90	1
91–100	0
101–110	−1

10. Look at the values in Table 2 for pulse rate while the subject was reclining. Locate the row in the data table below corresponding to this value. Now find the pulse rate 'Increase upon standing' in Table 2 to locate the correct column. Circle the appropriate cell where the selected row and column meet, and record the points in the fourth row of Table 4.

Reclining Pulse (Beats/min)	Points				
	Pulse Rate Increase upon Standing (# of Beats)				
	0–10	11–18	19–26	27–34	35–43
50–60	3	3	2	1	0
61–70	3	2	1	0	−1
71–80	3	2	0	−1	−2
81–90	2	1	−1	−2	−3
91–100	1	0	−2	−3	−3
101–110	0	−1	−3	−3	−3

Part 2: Step Test

This test will be performed on the stairs in the building. Bring these instructions and data sheets, a timer, and a writing instrument with you. What the test subject will be doing is a single exercise session for 20 seconds on the stairs. The other group members will then help measure the test subject's pulse rate constantly for the next two minutes to check his or her rate of recovery.

1. You will need a test subject (the same person used as the test subject in part 1), a timer, and a data recorder. Read the instructions for each person before you begin the test:
   a. Test subject: When the timer says to start, you will go down and up the first flight of stairs (using the rail for balance). After 20 seconds, the timer will have you stop. Try to finish the exercise period at the top of the stairs. Immediately find your pulse and start counting your heartbeats. The timer will have you call out the number you have counted at 15, 30, 60, 90, and 120 seconds after you stop. When the timer nods or points to you to indicate one of these time points, call out the number of beats you have counted. After each time point, immediately start the heart rate count over from one and wait for the next nod or point.
   b. Timer: Tell the subject when to start, and time him or her 20 seconds on the stairs. Warn the person when he or she has only five seconds left so he or she can end at the top of the stairs. After you have the subject stop, immediately have the subject start counting his or her pulse while you reset the timer and start again from zero. Point or nod to him or her to provide the number of beats that occurred during each of the time intervals. Stop the timer while the person calls out the number, then immediately resume as the test subject resumes counting.

You will ask for the number of beats after the first 15 seconds after exercise ('0–15 second interval'), then at 30 seconds ('16 –30 second interval'), then at 60 seconds ('31 –60 second interval'), 90 seconds ('61–90 second interval'), and 120 seconds ('91–120 second interval'). If the test subject doesn't start over from zero for each time interval, you may have to subtract the numbers to get the correct values.

c. Data recorder: Write down the numbers called out by the test subject at each time point for the correct time interval under 'Value from subject' in Table 3. Get the pulse rate in beats per minute by multiplying each value for intervals 15 seconds long by 4 and intervals 30 seconds long by 2.

2. We are really interested in how the pulse rate changed relative to the starting point: standing pulse rate. Write standing pulse rate from Table 2 into the top row of Table 3 for easy reference. Then, use this value to determine how the pulse rate changed for each interval compared to the standing pulse rate by subtracting the standing pulse rate from each pulse rate after exercise. Record the resulting values in the last column of Table 3. Positive values mean the pulse rate was faster than the standing pulse rate for an interval after exercise, while negative values indicate that the pulse rate after exercise was slower than the standing pulse rate for a particular interval.

Table 3—Change in pulse rate after a brief period of exercise. The value from the subject will be the raw numbers reported by the subject, which must be converted into beats per minute (60 seconds)

Pulse Rate	Value from Subject	Pulse Rate (Beats per Minute)	Change from Standing Pulse (Pulse Rate for Each Interval—Standing Pulse)
Standing	Take from Table 2 and record here ->		xxxxxxxxxxxxxxxx
0–15 Second Interval			
16–30 Second Interval			
31–60 Second Interval			
61–90 Second Interval			
91–120 Second Interval			

3. Look at Table 3 to find the pulse rate while the subject was standing. Locate the row in the data table below corresponding to this value. Now find the change in pulse rate for the 0–15 second interval from Table 3. Circle the point value where the row and column meet. Record the corresponding points in the fifth row of Table 4.

Standing Pulse Rate (Beats/min)	Points				
	Pulse Rate Increase for 0–15 Second Interval				
	0–10	11–20	21–30	31–40	41+
60–70	3	3	2	1	0
71–80	3	2	1	0	−1
81–90	3	2	1	−1	−2
91–100	2	1	0	−2	−3
101–110	1	0	−1	−3	−3
111–120	1	−1	−2	−3	−3
121–130	0	−2	−3	−3	−3
131–140	0	−3	−3	−3	−3

4. Look at the changes in pulse rate from Table 3. Find the first interval in which the change was either zero or negative, indicating full recovery from the exercise. Locate the corresponding row in the data table below. Circle the point value, and record the corresponding points in the sixth row of Table 4.

Seconds	Points
0–30	4
31–60	3
61–90	2
91–120	1
121+ and 1–10 beats above standing pulse rate	0
121+ and >10 beats above standing pulse rate	−1

Activity 3—Heart Rate of *Daphnia*

Materials needed for each group

Stereomicroscope
Timer
Cup or beaker
Depression well slide
Petri dish
Dropping pipette
Living *Daphnia magna*
Optional: slowing solution, such as *Detain 2*
(obtain just before it is used in your Petri dish): room temperature water, cold water, and crushed ice (or very icy water)

Introduction

Daphnia magna is an ectotherm. It is a small crustacean commonly found in freshwater ponds and lakes. It uses its large antennae-like oars, propelling its body rapidly forward as the antennae snap backward. The resulting jump-like movement gives *Daphnia* its common name, the water flea. *Daphnia* is highly transparent and all of its internal organs are visible. Its heartbeat can be observed with a stereomicroscope.

In this exercise, you will use *Daphnia* to study the effect of the environmental temperature on the heart rate of an ectotherm. Your teacher has set up water stations at different temperatures for you to use in your tests.

1. Obtain one concave–depression well slide. Place the slide with the concavity facing upward.
2. Add a *Daphnia* to the concavity of the slide, with enough culture fluid to almost fill the concavity.
3. Add one drop of slowing solution to the *Daphnia* and culture fluid. Slowing solution makes the fluid more viscous, or thick, so that you can more easily keep the *Daphnia* in your field of view.
4. Fill the Petri dish with water from one of the three temperatures, then place the lid (the larger piece of the Petri dish) on top. It doesn't matter which temperature you start with, as you will do the experiment for all three temperatures during this lab experience.
5. DO NOT turn on your microscope until you are ready to actually view the *Daphnia*! The volume of water on the microscope is so small that it is possible to overheat your *Daphnia* just by keeping the light trained on the specimen too long.
6. On your stereomicroscope, remove the circular disk on the stage (which is on the base for these microscopes). You will see a light beneath. Place your filled Petri dish in the slot vacated by the circular disk. Set the circular disk aside to put back on the microscope during clean-up.
7. Place the slide with the *Daphnia* on top of the lid of the Petri dish that is located on the microscope. Never put the slide in the water—the *Daphnia* will float away.
8. With the microscope off, allow the *Daphnia* to sit on top of the Petri dish for four minutes so the temperature has time to equalize.
9. Turn on the light UNDER the *Daphnia* on the stereomicroscope.
10. Quickly focus on the *Daphnia*, using the figure below to identify the position of the organism's heart. It is dorsal to the intestine. Do not confuse the motion of the antennae or the gills in the belly with the beating of the heart.

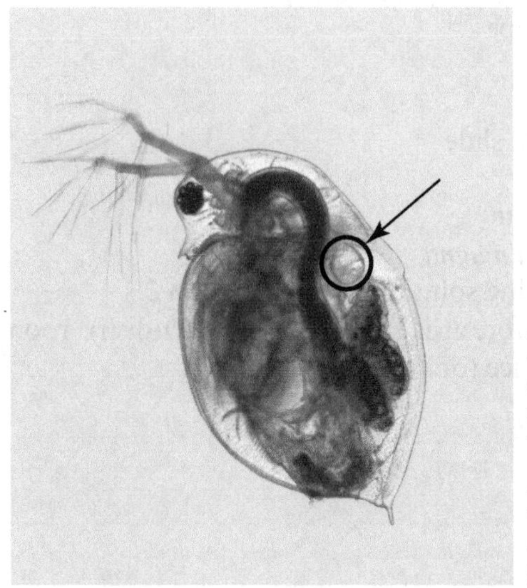

© Lebendkulturen.de/Shutterstock.com

11. Working quickly, one student should count heartbeats while another times a 10-second interval. Count the heartbeat for three separate 10-second intervals, and record the temperature and heartbeat data in Table 5.
12. After counting three intervals at a particular temperature, carefully remove the slide and set it aside.
13. Discard the water or ice in the Petri dish. Obtain a sample of water with a different temperature. Repeat steps 7 through 12, using water with a different temperature each time. Continue until you have tested water from all of the water stations.
14. Complete Table 5: Determine heart rate in beats per minute and average the resulting values, recording all numbers in Table 5.

Name ______________________ Date ______________

Physiology of the Circulatory System—Student Worksheet

DATA TABLES

Table 4—Scores for Physical Fitness Test

Test	Scores from Tables in Protocol
Blood pressure change upon standing	
Standing pulse rate	
Reclining pulse rate	
Pulse rate immediately upon standing	
Pulse rate increase after exercise	
Time to recovery after exercise	
Total Physical Fitness score (sum of all scores)	

Table 5—Heart Rate of *Daphnia*

Temperature	Heartbeats/10 sec			Heart Rate in Beats/min (Heartbeats/10 sec × 6)			Average Heart Rate (Beats/min)
Warm							
Cool							
Cold							

QUESTIONS

1. Using the total physical fitness score from Table 4 and the following chart, what is your relative fitness level? Do you think this is a fair test of physical fitness? Why or why not?

Total Score	Relative Fitness
18–17	Excellent
16–14	Good
13–8	Fair
7 or less	Poor

Name ______________________________ Date ______________

2. Calculate the following for at least one member of your group:
   What is your age in years? ________ years
   a. What is your maximum heart rate? (220 − age) ________ bpm
   b. What is your target heart rate for exercise?
      Low = (50% of maximum) ________ bpm
      High = (75% of maximum) ________ bpm

3. Compare how pulse rate changed compared to how blood pressure changed in Table 2. Do these results correlate with each other? Would you expect blood pressure and pulse rate to correlate? How and why? Include the major body structures and organs involved and describe their roles.

4. Why is it important that the subject's arm be at heart level when taking blood pressure measurements?

5. Consider two large mammals, a giraffe and a rhinoceros. If both animals were standing and relaxed, which would you expect to have the higher blood pressure? Explain your answer.

6. An astronaut's pulse rate on the day before launch is 65 beats per minute. After three weeks in orbit in zero gravity, the astronaut returns to the Earth. Would you expect the astronaut's pulse rate to be higher, lower, or the same as the day of launch after he or she returns to the Earth? Explain your answer.

7. Graph the temperature and heart rate data of *Daphnia.* Make sure to include axis labels and caption.
   a. The independent variable is ______________________.
   b. The dependent variable is ______________________.

Name ______________________________ Date ______________

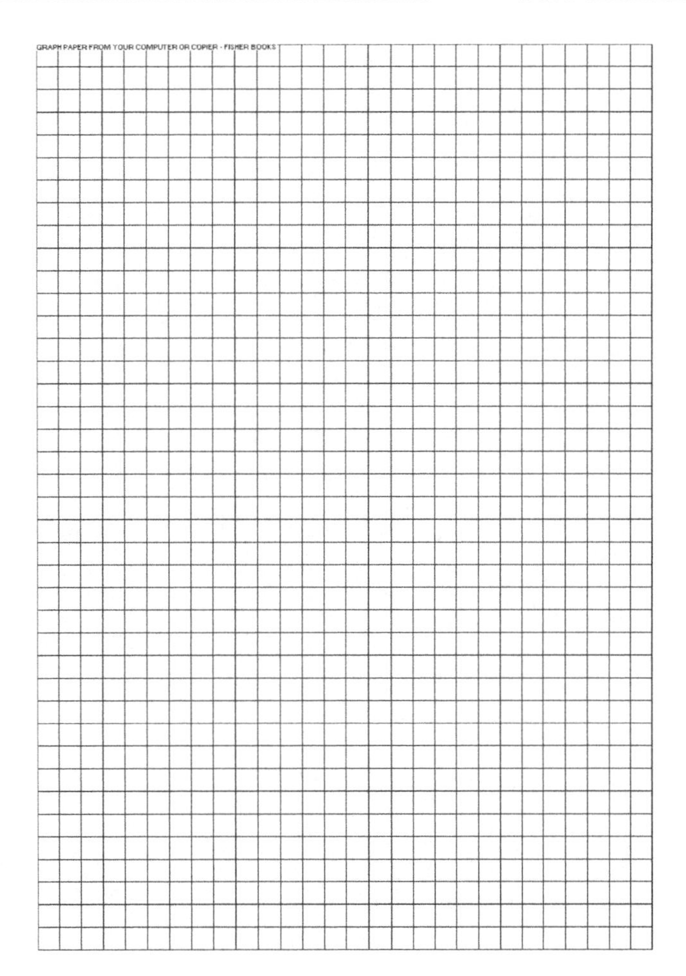

8. Write a hypothesis that this experiment is designed to test.

9. From your graph, summarize how temperature affected the heart rate of *Daphnia*.

10. Why does temperature change affect the heart rate of *Daphnia*?

11. Do you think that an endotherm's heart rate would change in the same way that we observed for *Daphnia*? Why or why not?

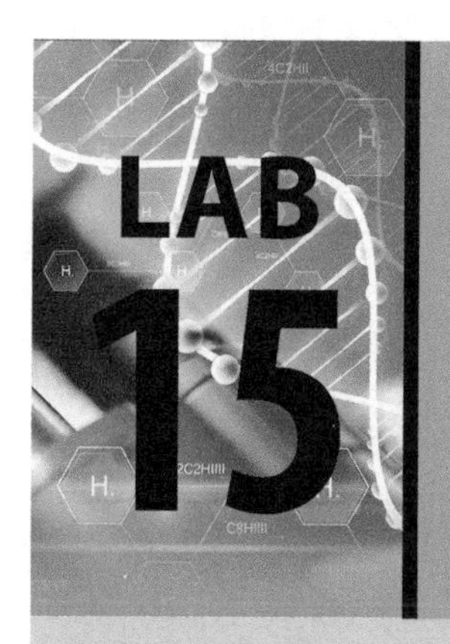

LAB 15 Simulated Disease Transmission

BACKGROUND

Infectious diseases are those which can be transmitted from a sick individual to an otherwise healthy individual. These diseases are caused by an infection from a pathogen that is now multiplying within the host's body while causing damage to it. Pathogens are any organisms or substances which cause disease by increasing in number within another organism and disrupting the normal functioning of that organism because of its growth. There are many bacteria, viruses, fungi, and eukaryotic parasites which fall into this category. These organisms are often referred to as microparasites because they are not visible to the naked eye. There can also be larger macroparasites which cause disease but are large enough to see with the naked eye; these include tapeworms, roundworms, and a variety of parasitic insect larva. There are many microorganisms growing on and within our body that do not interfere with our normal functioning and, in fact, can be beneficial. These beneficial symbiotic organisms are not considered pathogens.

Infectious diseases are very different than genetic diseases in that genetic diseases may be passed from parents to children, but never passed to someone other than to a direct offspring or descendent. Genetic diseases are not caused by a pathogen, but instead, specific disease alleles in the DNA of an organism.

Our body has a multi-leveled defense system known as our Immune System, but our defenses often need to encounter a specific pathogen before it recognizes it as being pathogenic. That is why we often get sick the first time we encounter a new pathogen, but can be resistant to it after that first illness. This is also the strategy behind vaccines. A vaccine is a collection of disease signature molecules (antigens), without having an active dangerous form of the pathogen present. The body is introduced to the signatures of the pathogen and marks it as dangerous so when a person is actually exposed to that pathogen, the immune system is already forewarned.

© kaktuzoid, 2013. Used under license from Shutterstock, Inc.

Infectious diseases are spread by the transfer of pathogens between individuals. This can occur through direct contact, fluid exchange (sexual and otherwise), indirect contact (touching the same surface that someone else had touched), aerosol

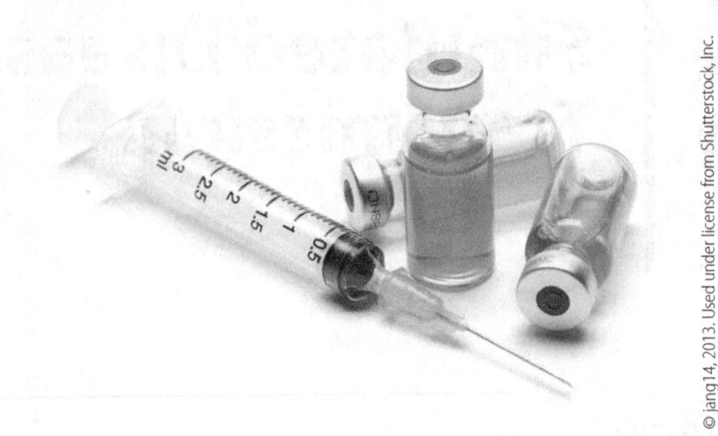

© jang14, 2013. Used under license from Shutterstock, Inc.

droplets released by a cough or sneeze, contaminated food or drink, and insect vectors, just to name a few methods. Studying diseases and how they travel through a population is a specific field of biology known as epidemiology. In today's class we will not only see the spread of a simulated disease through our class population, but will also have the ability to track the path of infection for that disease.

PROTOCOL

Activity: Transmission of a Simulated Disease

Materials needed for each individual

1 large snap-cap vial of simulated body fluid with ~30 mL of fluid
1 empty tube
1 transfer pipette

Shared components

Simulated Disease Indicator solution.

1. In this lab activity, all participants will be given a vial of simulated body fluids. Upon receiving the vial, each student needs to use the pipette to transfer 1 mL of solution from the large snap-cap tube to the empty tube. We will call this your reserve tube. Then discard the pipette, as it will not be used further in today's activity. (Note: Before the simulation starts, there will be one vial of simulated body fluids that start with our simulated disease. The disease and body fluids are not biological and are not actually pathogens, so there is no need to worry about actually getting sick in this particular lab activity. Through the course of this lab activity other individuals will acquire the simulated disease. If you acquire the simulated disease during this lab activity or happen to be the source of this simulated disease, it is in no way a comment on your character or personal hygiene. The original source was unlucky in choosing the infected vial.)
2. It is important to now seal the reserve tube and either initial it or place it in a location that you will know for sure that it is a sample of your simulated fluid before any exchanges take place. (Note: You do not want to mix up your reserve tube with that of another student. Also, whenever the cap is open on your large snap-cap tube, be

very careful not to knock over the tube. It can easily happen if you are not paying attention and it could cause the results for the entire class to be affected.)

3. **Rules for Fluid Exchange**
   A. Do not move on to fluid exchange until the instructor explains the rule for each round of exchange.
   B. There will be three rounds of fluid exchange and a different rule will precede each round of exchange.
   C. The rounds of exchange must always be simultaneous, meaning that everyone must find a partner and perform exchange round one before the entire group moves on to exchange round two.
   D. Each round of exchange must be with an individual that you did not exchange with in a previous round.
   E. Do not start the exchange of fluids until it is confirmed that everyone has found a legal partner for that round of exchange.
3. Fluid exchange in this activity consists of transferring ~10 mL of solution from the large snap-cap tube of one participant into the large snap-cap tube of another participant. Then capping the recipient's tube, mixing by inverting the tube a few times. Then transferring ~10 mL of solution back to the first, followed by another round of capping and mixing. (Note: Each tube should be starting with about 30 mL of solution, 10 mL are transferred to one tube, leaving one tube with only 20 mL and another with 40 mL. Cap and mix the tube with 40 mL, then transfer ~10 mL back to the tube that has 20 mL. This should return both tubes to 30 mL. In the fluid exchange, both participants start with 30 mL and both tubes should end with 30 mL.)
4. After completing the fluid exchange and recording the name of your exchange on the Student worksheet, return to your seat to await instructions on the next round of exchange. It is critical that you get the correct name for the person you exchanged with, and that you only do three exchanges during this exercise.
5. After completing three rounds of fluid exchange, every participant will return to his or her seat to be tested for this simulated disease. The class will be invited to the front or your instructor may come to you, one row at a time, for testing. Bring the large snap-cap tube and your student worksheet.
6. Testing involves adding seven drops of a blue "Simulated disease indicator solution" to the large snap-cap tube, then capping and mixing the tube. If the blue color remains, that is a negative response for this simulated disease. If the blue color disappears and the tube remains clear, that is a positive response to this simulated disease. You then need to record your name and the name of your three exchanges on the class data table on the board. The order of exchange is very important when you are writing on the board. For example, you are named Yelna, and you first exchanged with Loris, then with Potto, then with Lemur. If you tested positive, you would write the following on the board:
   Yelna
   1. Loris
   2. Potto
   3. Lemur

   That tells everyone that Yelna tested positive, and who she exchanged with in what order.
7. All participants will be tested and the results on the board should be recorded on your student worksheet.
8. We will then attempt to determine the order of disease transmission by ruling out individuals who could not be the source. We do this by looking at the exchanges. If someone had an exchange with a person who ended up testing negative at the end of the experiment, we know that at the time of the exchange, that person did not have the disease and so they could not be the source. They did not get the disease until later exchanges.

9. Once we have narrowed down the possibilities of who the source could be, those individuals must bring their reserve tube up with them for a final round of testing. A single drop will be added to the reserve tube of the potential sources and the one who has a clear tube after that round of testing is our source.
10. Cleanup for this lab involves discarding the reserve tubes and pipets and rinsing the large snap-cap tubes with water and leaving them open near the sink to allow them to dry.

Name ______________________________ Date ______________

Simulated Disease Transmission—Student Worksheet

QUESTIONS

1. Fill out the chart for your exchanges.

Exchange Round 1	Exchange Round 2	Exchange Round 3

2. Did you end up testing positive for the simulated disease?

3. Fill out the chart for the class data.

Name of Positive Individual	Exchange Round 1	Exchange Round 2	Exchange Round 3

4. Who was the source of our simulated disease?

5. Draw a flowchart of disease transmission, starting with our source and drawing a line to the names and showing who got infected for the first time in each round of exchange and from whom. Remember that you will need to draw an arrow between infected people in earlier rounds and the newly infected in each round, so the source will have infected a different person in rounds 1, 2, and 3, and will therefore have three lines coming from him or her. Those infected in round 1 will have infected two additional people, and those infected in round two would only have the chance to infect one person.

Source	Exchange Round 1	Exchange Round 2	Exchange Round 3

Name ______________________________ Date ______________

6. What causes infectious diseases?

7. How many people were infected in class after three rounds of exchange? It is possible that some infected people exchanged with other infected people in some rounds, reducing the number of newly infected. Figure out the maximum number of people who could have been infected during our simulation. Exclude the source, since he or she was not infected during our exchanges. Did we have the maximum number of people infected?

8. Why were there names on the board of individuals who tested negative?

9. How can one person spread a disease to so many other people?

10. What percentage of the participants in our class tested positive for the disease?

11. What would have happened if we would have had a fourth round of exchange? Be as quantitative as possible, indicating how many new people could have been infected in a fourth round given our results.

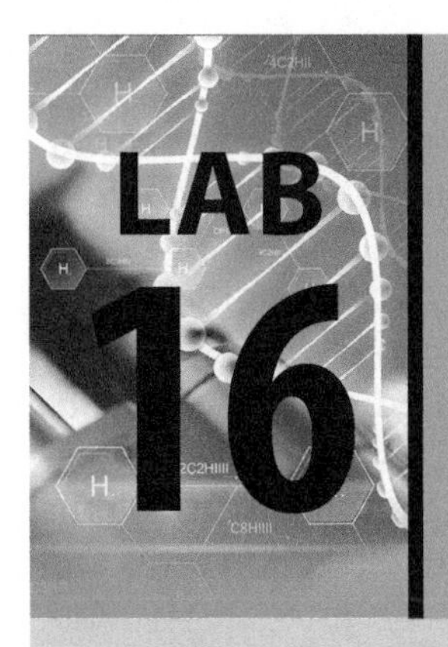

LAB 16 Fetal Pig Dissection Instructions

BACKGROUND

There are many similarities between the anatomy of humans and pigs. There are several reasons for this. Both humans and pigs are vertebrates, meaning that we have backbones and skulls. We are both mammals, meaning that we are warm-blooded, have a four chambered heart, closed circulatory system, hair, and mammary glands. We are also both omnivores, meaning that our teeth and digestive systems are capable of regularly consuming both plant and animal material. Our reproductive systems are similar in that the female gonads (the ovaries) are an internal organ while the male gonads (the testes) are external structures contained within the scrotum. The size and cost of dissecting fetal pigs allows first semester biology students access to dissections that would not be feasible on human cadavers until later biology courses. The purpose of today's lab is to get an introduction to dissection tools and dissection techniques along with identifying some of the common structures of the pig's internal anatomy.

In order to understand the instructions for the dissection, there are certain anatomical terms that must be learned. These terms are used to direct and navigate the dissector as they peruse the internal organs of the fetal pig.

Dorsal—The back surface, toward the spine
Ventral—The front surface, toward the abdomen and stomach
Anterior—Above, toward the head
Posterior—Below, toward the tail
Lateral—Away from the center
Medial—Toward the center
Superficial—Toward or along the surface
Deep—Substantially below the surface
Longitudinal—A vertical line that runs anterior and posterior
Horizontal—A line that runs right and left
Dextral—To the right of
Sinistral—To the left of

The Dissection tools which will be used in this dissection include the following:

Scalpel—Bladed dissection tool used to cut tissue.
Scissors—Dissection tool with paired blades which can be used to cut tissue without damaging underlying tissues.

Forceps—Tweezers, used to hold on to and lift small objects and manipulate the specimen.
Probes—Blunt dissection tool used to move portions of the specimen in order to get a better view of underlying structures.
Dissecting Needles—Small sharpened dissection tools used to pin specific sections of the specimen to the dissecting tray.
Gloved Hands (not included in the box of dissection tools)—located at the distal ends of your arms, will be used to manipulate the dissection tools as well as directly manipulating the specimen. Do not be afraid to pick up the fetal pig or directly touch the tissues of the pig. In general, your hands are more dexterous than dissection tools and can allow for better viewing of the structures of the fetal pig.

We will be looking at several different organ systems in today's dissection, including the digestive system, the circulatory system, the respiratory system, and the reproductive system. Motivated students may also observe additional organ systems (nervous system, etc.) as class time allows.

PROTOCOL

Observing the External Anatomy

1. Obtain a dissecting tray, fetal pig, gloves for all group members that will be touching the pig and the tools, and a complete set of the dissecting tools that you will need.
2. The preserving solution is very strong so open the bag and rinse off the pig in the sink along with pouring out of the residual preserving solution (it is biodegradible and nontoxic when diluted, so rinse with an abundance of water, then return to your groups working area with your washed pig on the tray).
3. Make a thorough examination of your pig, including its facial features, limbs, and gender (male pigs have scrotum anterior to the anus and a urogenital opening just posterior

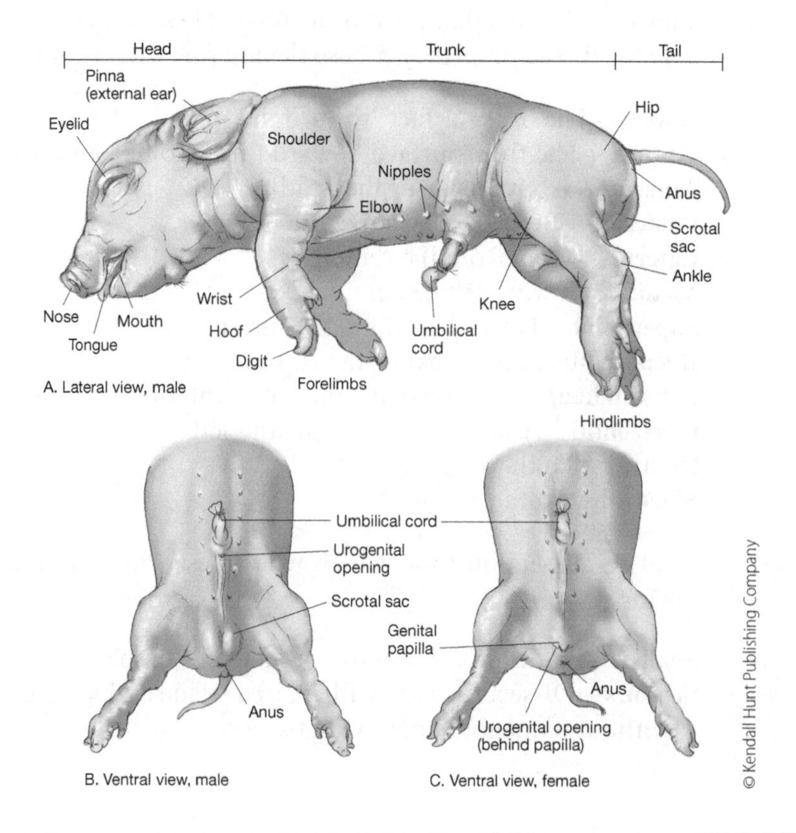

© Kendall Hunt Publishing Company

of the umbilical cord, female pigs have urogenital opening anterior of the anus that is enclosed by small folds of skin called labia that form the genital papilla, a small projection. If your pig has a visible protrusion near the anus, it is likely a female.)

4. Once the gender of your pig has been determined, decide as a group on the name of your pig and record its name on your worksheet.
5. Place the pig with its ventral side up to prepare to open the abdominal and thoracic cavities.

Opening the Abdominal and Thoracic Cavities

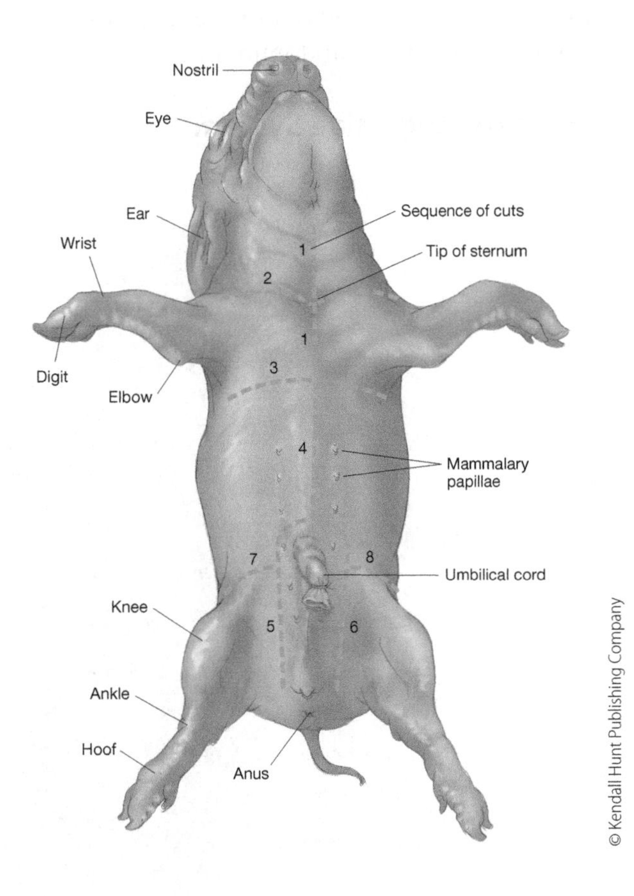

1. Using the scalpel, make a superficial longitudinal incision starting below the chin and cutting posteriorly towards and then across the surface of the rib cage. This first incision should be cutting through the skin and muscle, but not yet the bones of the rib cage.
2. Pull the skin and muscle laterally to reveal the sternum of the rib cage. After the surface of the rib cage is exposed, use the scissors to cut the bones of the rib cage, taking care to not smash and destroy the structures of the deeper tissues. Make sure that this cut is lateral to the sternum, as the breastbone is harder to cut through than the neighboring ribs. Continue cutting through the skin, ribs, and muscles until you reach the posterior edge of the rib cage and the diaphragm (an internal, horizontally-aligned muscular sheet used during breathing).
3. Cut a horizontal line anteriorly and posteriorly to the forelimbs, along the edges of the ribcage. Move the forelimbs laterally to obtain a better view. Two longitudinal incisions on the lateral edges of the ribcage can be performed to remove the surface of

the rib cage from obstructing the view. Separate the edge of the diaphragm from the abdominal wall if necessary.

4. Continue the center longitudinal incision from the diaphragm to a location just anterior of the umbilical cord.
5. From the umbilical cord, continue the incision dextrally and sinistrally in a posterior manner to the genital region. There is an umbilical vein that is anterior to the umbilical cord that may still be connecting the umbilical cord to the anterior portion of the abdomen. After this vein is severed, the entire skin flap containing the umbilical cord can be folded posteriorly to increase the view of the abdominal organs.
6. Make a horizontal incision on the body wall just anterior of the hind limbs on both the right and the left hand side. The skin and muscle along the sides of the abdominal cavity can now be moved out of the way or removed using a longitudinal incision on the lateral edges of the abdominal wall.

Organs of the Digestive System

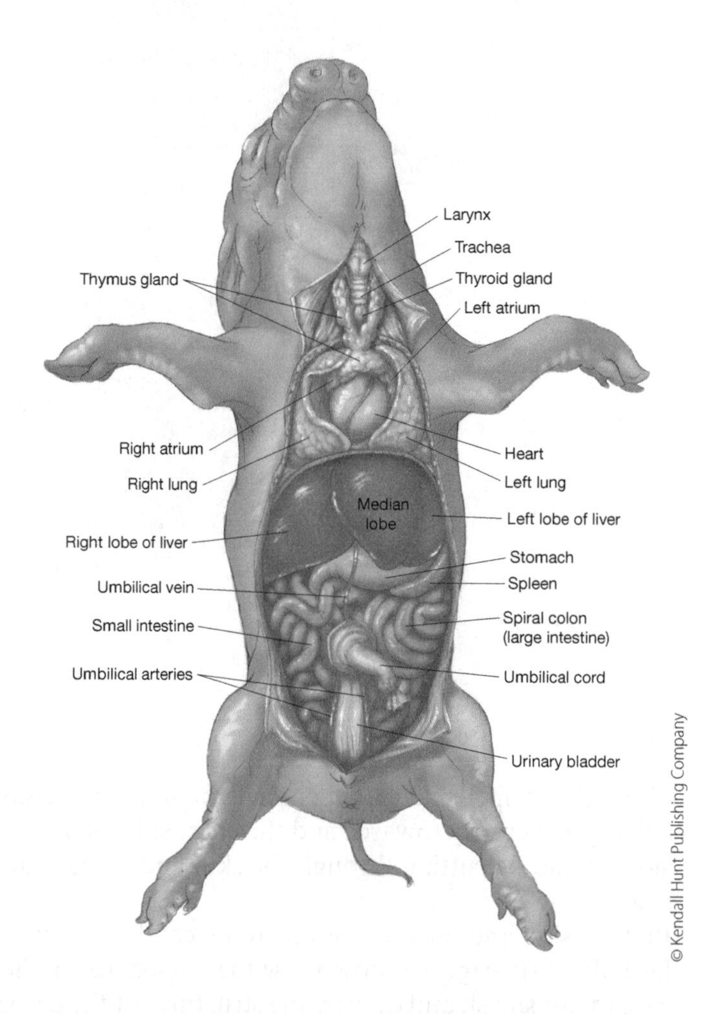

© Kendall Hunt Publishing Company

1. The organs of the digestive system in pigs are very similar in shape and orientation to the digestive system in humans. Begin your observation by identifying the stomach, small intestines, and large intestines. These structures will all be found below the diaphragm.
2. From the stomach, search anteriorly to find the esophagus and posteriorly to find the duodenum (first part of the small intestine). Near where the stomach and duodenum meet, deeper into the abdominal cavity, there will be a grainy organ connected to

(and possibly surrounded by) the mesenteries known as the pancreas. This will be found between the duodenum and the spleen, behind the stomach.

3. The liver will be a dark prominent multi-lobed organ below the diaphragm and near the stomach; count the number of lobes of the liver present in your pig and see if you can identify the gallbladder, usually located between a few of the lobes of the liver and transparent or greenish in color.
4. The small intestines will transition to the large intestines. Continue to follow this path to the large intestines near the anus. Cutting the lower portion of the colon and the most anterior portion of the esophagus posterior of the diaphragm will allow this entire portion of the alimentary canal to be removed (with the removal of the connective tissue holding it in place), providing a view of the back wall of the pig's abdomen, including the kidneys. The liver may also be removed by cutting the anterior connective tissue holding it in place.
5. The kidneys will be revealed, attached to the dorsal wall of the abdomen. Some connective tissue may need to be removed in order to clearly view the kidneys.
6. Unwind the small intestines and large intestines by cutting them loose from the mesenteries that are holding them, and compare their relative lengths and diameters.

Organs of the Circulatory and Respiratory Systems

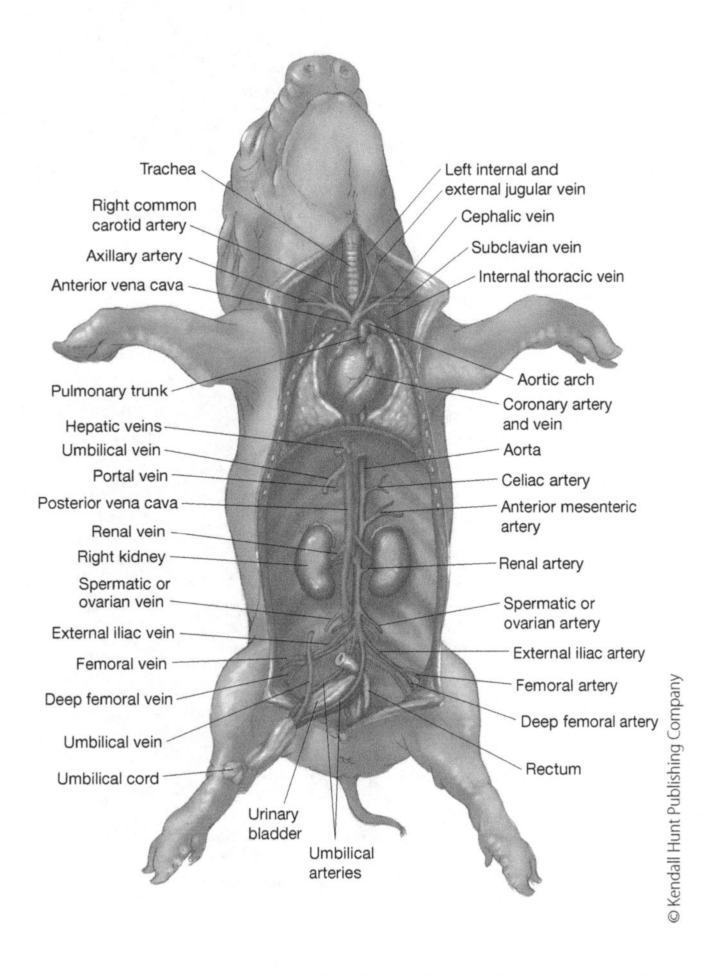

© Kendall Hunt Publishing Company

1. The internal organs of the chest cavity include the major organs of the circulatory and respiratory systems. Upon opening the chest cavity, identify the heart and the lungs. The heart may likely be located under a layer of connective tissue known as the pericardium. This pericardium may be removed to allow for a clearer view of the heart.

2. Like humans, the heart of the pig has four chambers; two upper chambers known as the atria and two lower chambers known as the ventricles. See if you can identify where these structures are found in the heart and draw an image of the external view of the fetal pig's heart.
3. Identify the lungs and the cartilaginous trachea (windpipe). This may require the moving of or removal of the thymus gland.

Organs of the Reproductive System

1. Your pig will be male or female, which you had established with the observation of the external anatomy. The internal reproductive anatomy will be different for the two sexes.

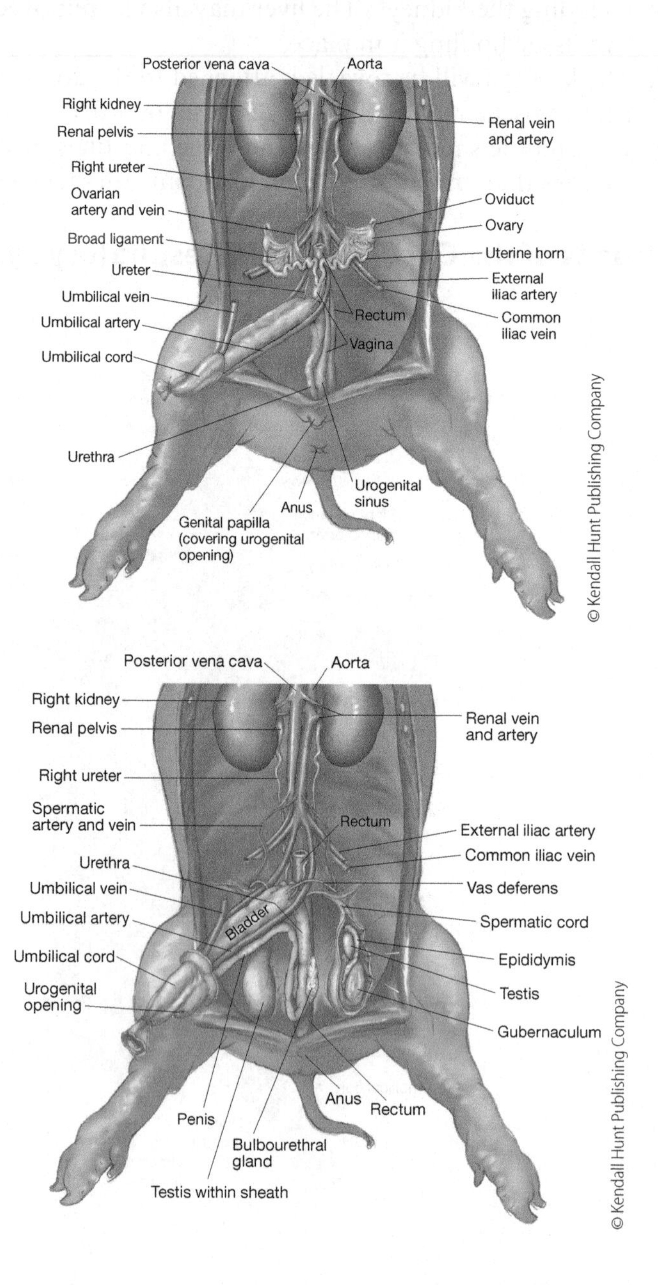

2. If you have a female pig, identify the urinary bladder, vagina, uterine horns, and ovaries.
3. If you have a male pig, using a scalpel or pair of scissors, open one of the sheaths surrounding a testis and identify the testis, epididymis, penis, and urinary bladder.

4. After completing your observation of the organs of the reproductive system, find a lab group with a pig that is of a different gender than your own and observe the above listed structures.

Clean Up

1. After completing the directed dissections and any additional investigatory dissections of interest, the pig and any piggy pieces should be disposed of in the red biohazard waste bag near the front of the room.
2. The dissection trays and tools should be rinsed with water and dried, then returned to the area where they were obtained at the beginning of the dissection.
3. The dissection area should be wiped down with a damp paper towel or disinfecting wipe with special attention taken to wipe up any residue that may have been left behind.

Name ______________________ Date ______________

Fetal Pig Dissection—Student Data Sheet

1. What is the name and gender of your pig?

2. Describe the ears of your specimen. Be sure to include details such as color, shape, and texture in your description.

3. Draw a picture of the stomach (organ) in the space provided and write a description of it.

4. Describe the pancreas of your specimen in your own words.

5. How many lobes made up the liver of your pig? Were you able to find the gallbladder in your specimen? If yes, describe it, if not, explain where you looked to find it.

6. Why are there so many blood vessels in the mesenteries connected to the intestines? (Hint: What is happening in the small and large intestines?)

7. Draw a picture of one of the kidneys in the space provided and write a description of it.

Name ______________________________ Date ______________

8. Which was longer, the small intestine or the large intestine? Which was wider?

9. Draw a picture of the external view of the heart in the space provided, label the chambers.

10. Draw a picture of the gonads of your pig. Label the gonads with their gender specific name.